THE SASQUATCH PEOPLE

AND THEIR INTERDIMENSIONAL CONNECTION

The Sasquatch People and Their Interdimensional Connection

Published by Comanche Spirit Publishing

Printed in USA

ISBN: 978-0-9833695-3-0

Books may be ordered by contacting the
author at Sasquatchpeople@hotmail.com
or 1-425-844-8409

www.Sasquatchpeople.com

Another book by Kewaunee Lapseritis
The Psychic Sasquatch and Their UFO Connection

Editing, text design, and typography by Carol Hiltner

Cover art and the drawings at the beginning of each chapter
by Jesse D'Angelo

THE SASQUATCH PEOPLE

AND THEIR INTERDIMENSIONAL CONNECTION

KEWAUNEE LAPSERITIS M.S.

PREFACE
CHRISTOPHER L. MURPHY

FOREWORD
LEE TRIPPETT

DEDICATION

This book is dedicated to all the Sasquatch contactees who have had the courage to step out of the shadows and into the light. Without their openness and sharing, this book would not have been written.

Also, this book is dedicated to Maris Kundzins (1942–2010) whose own interdimensional journey happened way too soon. Maris, just know that we miss you, and that Sheila, Mario, and I will never stop loving you.

Maris Kundzins (1942–2010)

TABLE OF CONTENTS

"Apes can walk on two legs, but not with the strides and gait the Patterson Bigfoot uses. That's a human trait."

Dr. Esteban Sarmiento, Primatologist
Previously of the American Museum
of Natural History, New York

"The exact contrary of what is generally believed is often the truth."

Jean de la Bruyere

"Using advanced technology, when I freeze and enlarge a frame from the Patterson film, it's plain to see that the creature has lips. Only humans have lips, not anthropoid apes."

M. K. Davis, Researcher

ACKNOWLEDGMENTS

First to my dear friends and associates Lee and Marlys Trippett with my deepest gratitude for their unwavering support over the years;

Many thanks to my good friend, associate, and staunch supporter Mike Wisotzke, who firmly believes in me because he was a witness to some of these anomalous events I effected and manifested at times with the help of the Sasquatch people;

Much love to Rusty and Irene Axtell for their ongoing support;

My warmest appreciation to Jeremy Lynes for his brotherly help and for his always being there to assist me in the kindest way;

May God richly bless Mary Rau for her spiritual support and generosity. You are certainly a most wonderful and loving friend;

I am most grateful to my spiritual brother Donald Peterson for his friendship and research assistance; also for the support from his wonderful companion Kathleen Ryan;

My kindest regards to forensic scientist Dr. Kenneth Siegesmund for his support analyzing my physical evidence for the last 30 years;

Thank you kindly, John Allen, for believing in my work and always encouraging me to carry on. You are also my spiritual brother;

My love and blessings to Professor John Boatman for his superb contribution to this book as an Ojibway First Nation person. Recently, the eagle guided him home over the Great Divide. He will be missed by many;

Jack and Audrey Jurkonie's friendship and moral support over the years has meant everything to me. Thank you and God bless you both;

One of my greatest supporters and a noble friend, Russ Rusby, has been there for me every step of the way and is one of the most generous individuals I have every met! Thank you with all my heart.

My utmost appreciation and blessings to experiencer Susan Treibs for her spiritual support over the years;

Dr. R. Leo Sprinkle must be mentioned here. My thanks to him over the last 25 years of friendship and for his support in taking the time to critique the manuscript to get it polished up. Many blessings to you and your family;

On an expedition to the Oregon Cascades in 1980, I had the pleasure of meeting Phil and Wendy Singleton, and we have been the best of friends ever since. I am most grateful for their love and support while conducting field research in that part of the country;

Many thanks to Elaine Freeman for her love, kindness, and generosity. Your caring heart is most appreciated;

My profound gratitude to my editor and designer Carol Hiltner, whose professionalism shines through at every level;

With my deepest appreciation to Jesse D'Angelo, whose professional talent and exquisite artwork have brought my writing to life;

Many, many thanks to my typist Terry Burch for her unending patience and expertise in getting everything right. May God graciously bless you and your family;

I want to express my gratitude to Glenn Franco Simmons for his photography of the Redwood forest used for the background on the cover of the book;

And lastly, it has been a blessing, Maris Kundzins, to have you for a faithful friend for 46 years! It's been an interesting adventure. Thank you for the photography you contributed in the book and for your constant support. As a Zen Master said, "Every step of the journey *is* the journey."

PREFACE
BY CHRISTOPHER L. MURPHY

The depiction of strange ape-like creatures in the history of North America predates European settlement. Native people show such creatures in pictographs (painted images) and in petroglyphs (stone etching). In more recent times, they were depicted in woodcarvings (totem poles and ceremonial masks). Stories of such creatures abound in Native literature, and always state that such beings were "spiritual" or "paranormal" in nature. In other words, they possessed powers beyond those of normal humans.

Although Native cultures also include numerous other "spirit" creatures, remarkably, only the one with ape-like features found its way into non-Native culture. In about 1925, this creature became known to non-Natives in Canada as the "Sasquatch," and then in the 1950s the familiar term "Bigfoot" became its common name in the United States. This name, of course, was derived from the large human-like footprints that have been found (and continue to be found) in wilderness and rural areas.

The main reason non-Native people have professed a belief in Sasquatch/Bigfoot is due to physical sightings of the creature or other evidence of its presence, of which there have been at least 2,500 incidents over the last 100 years or so. At this time, it is estimated that there are at least 400 sightings/incidents per year throughout the United States and Canada. Perhaps the earliest US newspaper report on what appears to have been this creature was published in *The Exeter Watchman*, New York, in 1818. It reads as follows:

> **Another Wonder**
> *Sackets, Harbor, New York, August 30, 1818*
> Reports says, that in the vicinity of Ellisburgh, was seen on the 30th Ult. By a gentleman of unquestionable veracity, an animal resembling the Wild Man of the Woods. It is stated that he came from the woods within a few rods of this gentleman—that he stood and looked at him and then took his flight in a direction which gave a perfect view of him for some time. He is described as bending forward when

running—hairy, and the heel of the foot narrow, spreading at the toes. Hundreds of persons have been in pursuit for several days, but nothing is heard or seen of him.

The frequent and positive manner in which this story comes, induces us to believe it. We wish not to impeach the veracity of this highly favored gentleman, yet it is proper that such naturally improbable accounts should be established by the mouth of at least two direct eyewitnesses to entitle them to credit.

Since that time, thousands of newspaper articles on Sasquatch sightings and other related evidence have been published. Although many are certainly the result of overactive imaginations, misidentifications, and hoaxes, it defies logic to dismiss them all as such.

Serious research into the question of actual Sasquatch existence started in the 1950s. It was (and continues to be) reasoned that the vast expanse of North America could reasonably conceal a creature of this nature, and it was thought that it would only be a matter of time before a specimen (physical body, or part thereof) could be placed before men of science. The filming in 1967 (Patterson/Gimlin film) of what is believed by many to be a Sasquatch bolstered belief in its

Bob Gimlin at Kewaunee's cabin. Bob was witness to the filming of a Sasquatch at Bluff Creek, California on October 20, 1967.

existence and greatly increased hope that one would be captured, killed, or at least again photographed with equal or better clarity.

Remarkably, such has not been the case. There have been many associated tangible "findings" (alleged footprints, body prints, hair, feces, sound recordings, and so forth), but nothing that has fully satisfied the scientific community that the creature exists. Nevertheless, through the efforts of Dr. Jeff Meldrum, the world of science recognizes that something unknown makes large human-like footprints.

Some years ago, I concluded that the Sasquatch/Bigfoot issue was like a jigsaw puzzle with essential parts missing. It appeared to me that these parts might be beyond normal, rational (conventional) thinking. In other words, one might need to go outside the restrictions of physical science to find answers. Ironically, Native people had "been there" long before non-Natives set foot on their continent.

My first in-depth exposure to such thinking was in meeting Kewaunee Lapseritis in the mid-1990s. Kewaunee sent me his first book *The Psychic Sasquatch and Their UFO Connection*, and over the years we have kept in touch and established a good friendship. He has confided many things to me and sent me numerous photo-

Author/researcher Chris Murphy and Kewaunee Lapseritis at a Sasquatch Symposium in Bellingham

graphs of the research he has continued to undertake. I told him I was only really interested in "physical" things—having concluded that he had just as much chance of finding actual Sasquatch relics or artifacts as anyone else.

We have discussed at length his unusual experiences, and although, like the Sasquatch itself, neither he nor I can put these "on the table" and physically analyze them, I firmly believe in Kewaunee's honesty and sincerity and think he has certainly made important discoveries. The fact that he is not alone in this field is also highly important. Numerous other people say they have had essentially the same experiences, and like regular Sasquatch sightings, one gets to a point where quantity starts to say something. As John Green says, if just one of the thousands of Sasquatch sightings can be proven beyond a doubt, then there can be no doubt that the creature exists.

I think the most important thing scientists and people in general can do with regard to the paranormal in all aspects is to listen. There have been many cases in history where claims have been made and cast aside as ridiculous or impossible, only to later come to light as totally correct—some have "changed the world." Could the Sasquatch be of the nature that Kewaunee tells us? Of course it could. What we know of the universe has hardly touched what there is to know.

Kewaunee's unique spiritual and eclectic field approach to this research may not meet with the approval of his peers and mainstream scientists because it goes beyond the unidimensional biological sciences into the little-known realm of quantum physics, psychic phenomena, and the outer limits of the intangible Sasquatch world. Nevertheless, at least Kewaunee has the courage to express the truth as he has come to understand it, and is willing to share his knowledge in this unique landmark book. For certain, readers will become acutely aware that there is possibly much more to the Sasquatch/Bigfoot issue than they could have ever imagined.

FOREWORD
BY LEE TRIPPETT

Our family came across first-hand news of unusual and remarkable events in the early 1960s. My Dad's brother was in Happy Camp, California, when there were reports of hair-covered giants. And when he shared these events with us in Eugene, Oregon, our family became enamored with intrigue and potential adventure.

This was the beginning of a lifetime of searching for reasonable explanations of phenomena that were totally new in all the knowledge of what we thought was our extensive awareness. How is it possible for a giant to survive and keep out of sight from hundreds of trackers, loggers, hunters, and miners? This is the one vital question that has remained throughout my nearly five decades of often active interest. This is still the one question that has opened many doors to a totally unexpected range of awareness.

Why five decades? Because the amount, quality, and character of the evidence kept expanding. Thousands of reports over a wide range of time, geography, and witness credentials show a consistent pattern. The more we interviewed witnesses, and heard or read of reports, the more compelling was our need to find a reasonable answer. Being schooled in advanced hard science that has inappropriately relied on only replication, there was a prevailing drive to find support for those witnesses who needed understanding for their sometimes earth-shaking experiences.

Through the 1960s our family came to know major players in the field of research in the Pacific Northwest. We came to understand the impossibility of looking behind millions of truly big trees all at once and during the night. This required a unique research tool because it was becoming obvious that traditional advanced technology was not working.

I reluctantly took the suggestion of various "proven intuitives" to camp alone out in a very remote area and wait. No guns, cameras, or hidden motives allowed. (It took years to fully understand and appreciate the reason for this style of approach.) So in the late 1970s, when my employer gave total freedom of hours to the entire staff,

I spent two consecutive nights during the full moon for ten consecutive months. As always, like many times before and after, I paid very close attention to details of the environment and eventually noticed things out of order. I realized that I was being watched by something very powerful and far more in tune with nature than I could have ever imagined.

Being a bit slow to catch on, it wasn't until the 1980s and '90s that I started paying more attention to subtle events, energies, the feelings of witnesses, and Native American stories. My slowness to "catch on" to what could be happening was due, in large part, to early training in religion and hard science that required "reality" to fit within established scriptural doctrine or objective measurable parameters.

The really big shift in expanded awareness about the nature of these hair-covered giants occurred slowly with four events. One was a personal meeting with Kewaunee, the author of this book. Another was meeting with witnesses who had had multiple sightings and telepathic conversations with these giants, and third, the remarkable discovery in quantum science of an "omnipresent, omniscience, and omnipotent consciousness" by my peers in, of all places, the field of physics. The fourth was a long-standing interest in alternative health. It seems that there are now a number of health care practitioners who can have a very relaxed attitude with focused coherent intent to bring about healing miracles with the approval of a consenting client.

This is a direct confirmation of ancient healing practice and the recent discovery of quantum science. It is essential to not let the profound implications of this go by. It is these words and thoughts that have provided meaning far beyond the typical modern world-mind. It is these words and thoughts that provide the mental tool that allows the giants to get along without the need of material artifacts. It is these words and thoughts that have provided wide ranging extra-ordinary powers to affect matter and energy in ways that we describe as paranormal or psychic events. It is also these words and thoughts that were taught 2,000 years ago.

However, those concepts are now translated into something meaning a deep unshakable confidence in faith. Our familiar prevailing attitudes and thought patterns have far more significance on our life experience than we realize. Our potential is unlimited for access to real freedom. In other words, we need not be limited by energy, time, space, and form. We all have a lot to learn from these hair-covered giants.

How much expansion can our consciousness take?

How much energy or influence can a disciplined mind affect?

The search for a reasonable answer to profound phenomena has led to considerable study in the forefront of religion and science. The ride has been a wild, entertaining, and exciting learning adventure over the past 45+ years.

Within the following pages you will find all the important necessary arguments, first-hand personal experience, and scientific leads for evidence that we and other humanoid forms of this incredible universe have talents and capabilities that far transcend the limited information that can be commonly found in all the textbooks of our society. Quantum science and the remarkable personal discoveries of the author will open the door to many exciting awakenings.

Best Wishes!
Lee Trippett
Eugene, Oregon
April 2010

Treat the Earth and all that dwell thereon with respect.
Remain close to the Great Spirit;
Show great respect for your fellow beings;
Work together for the benefit of all Mankind;
Give assistance and kindness wherever needed;
Do what you know to be right;
Look after the well being of mind and body;
Dedicate a share of your efforts to the greater good;
Be truthful and honest at all times;
Take full responsibility for your actions.

Amerindian philosophy;
the Sasquatch people and
Ancient Ones' spiritual beliefs

INTRODUCTION

"Oh, you're a contactee?" The few who have been sought out by the psychic Sasquatch are indeed honored, because the elusive giants will only speak or intentionally show themselves to humans who have open hearts. Contactees are at the forefront of a socio-psychological evolution that is being instigated by these highly advanced nature beings whom I call the Sasquatch people.

In recent years amongst Sasquatch researchers, however, the word "contactee" has become a dirty word—a word with connotations synonymous with pseudoscience, misinformation, subjective data collecting, or worse yet, the "lunatic fringe." The reality is that more and more open-hearted people are being contacted by the psychic Sasquatch, and I have become a key facilitator—helping witnesses to expand their spiritual consciousness by interacting with these evolved beings. Ironically, it is the contactees who are the real experts, because via telepathic communication they have been enlightened as to who and what these forest giants truly are.

I unexpectedly became a contactee in 1979 when both a Sasquatch and ET telepathically spoke to me. It was a very startling experience, but has been an ongoing process ever since. Because of my scientific background, I struggled for two years with the psychic aspects of this phenomenon until I finally stopped my denial and accepted this new reality. My life has not been the same since the initial contact.

I have now been conducting Sasquatch research for 55 years. My struggle with a resulting spiritual transformation has been sorely misunderstood by the community of researchers who believe that procuring a dead body, bones, or clear photographic evidence is what this phenomenon is about. But there are no monsters here for the taking!

Such perceptions come from people who have never seen a Sasquatch or amicably interacted with one via mental telepathy. How can people proclaim themselves "experts" in this field without fully experiencing the subject that they are investigating and supposedly studying? Those who have been most critical of contactees have the least to show for their efforts! Sadly, this field has few bona fide scientists and an abundance of armchair critics, computer geeks, and weekend warriors, when long term field work is needed—in the Jane Goodall tradition. However, the real breakthroughs are happening to the consciously aware people who have the patience and lack of fear to spend weeks or months alone in the wilderness without a gun or camera. Despite the adolescent attitude of some scientists toward the unknown, especially involving psi, UFOs, and other elusive phenomena, many sagacious scientists realize that quantum physics and psychic phenomena are synonymous—perhaps born of the same crucible.

Statistically, as of November 2010, I have documented exactly 187 experiencers who can objectify the reality of a psychic Sasquatch. In the following pages I share my personal story and those of witnesses as they unfolded. I have made no effort to "scientifically" prove these anomalous experiences, because the scientific paradigm to measure reality is incomplete and nonholistic, and thus the subject is viewed in distortion. So it is up to the reader to decide what is real.

The reader may wonder what my background and scientific methods are, that have led me into such controversial viewpoints. I was raised hunting, fishing, and trapping as a boy in the 1950s. Later, I backpacked around the world to over 40 countries by age 25. Ninety-nine percent of the time I conduct my field work alone. I conducted an ethnographic study, learning from Amazonian Indians, and later studied herbology with the Great Lake Woodland Indians. These experiences certainly helped me to be confident and independent in the field. But it's what I hold in my heart that counts the most, and I still have to work at it by trying always to be a better person in service of God and humanity. That's most important to me—not some scientific physical proof that in the end may alienate the giants from me. We have built a trust and I am fortunate to be learning from them.

I study botany—both medicinal and edible plants in the areas I visit. I read, write, pray, meditate, and telepath; in other words, I relax and merge with nature. Taking on some of the qualities of a Sasquatch in a non-threatening way has proven successful. I go out in the forest for one week, three weeks, sometimes more—once up to five months in one place to make contact. I let them come to me as they feel comfortable after monitoring my behavior for a time. Because of their powerful psychic networking to other clans and tribes *around the globe*, they know who I am before I even enter the area. This is fact. I telepath to the one near my home before I leave, asking for permission from the others out of respect. I certainly wouldn't want anyone camping on my front lawn without first asking my permission.

The beings say they have wisdom to share, but our inflated egos are blinding us from the truth. From time to time, when I let my own ego interfere with this sensitive research, they have stated to interspecies communicator Kathleen Jones that I need to get out of my head and back into my heart while letting go of negative emotions. This message is really for everyone. That's not always easy for me, but I try. I am only scratching the surface of a whole new area of anthropology and quantum science that the mainstream conservatives have been ignoring.

There is a vital need to change the scientific paradigm, because it does not include the aspect of quantum physics, which would change the way we view our world concerning Sasquatch. After 32 years it's still fascinating to me to experience the now-very-normal paranormal phenomena. I have my own proof that Sasquatch interdimensionalism exists. If I tried obtaining physical proof, I would be regressing. For I cannot unlearn how to swim!

With this book some may feel that I stretch the present scientific model, yet it is time that the world be properly informed of this new para-physical paradigm. New areas of science are being discovered all the time and this is one of them. Whether recognized or not, this phenomenon is here to stay. In fact, the contact and telepathy are accelerating all over the country, perhaps the world!

Though greatly criticized for not publicly revealing names and addresses of contactees/interviewees, I feel it is not fair to subject them to certain ridicule and scrutiny by those who don't understand the complexity of this spiritual phenomenon. I make no apologies for the fact that my book is largely anecdotal. As a scientist, I took the limits of scientific method as far as they would go, always objectively following wherever the evidence would lead me.

It is important to note that the Sasquatch as a people, with numerous clans and tribes throughout the North American continent, have asked me to write this book about them in order to directly affect societal awareness of them. Their purpose is not only to make the public aware of their clandestine existence, but to impress upon the outer world that *many* other types of beings also exist on this planet—some from outer space, some from inner space.

As a responsible researcher, I feel my discoveries need to be shared with the public. I am the first academician to experience and document the dynamics of mental telepathy and psychic healings, and to witness interdimensionalism in the quantum realm of the Sasquatch people. I am proud of my achievements in presenting a book based on integrity, perseverance, and love.

None of this research was compromised to please the skeptics. When I asked the Ancient Ones how much of what they told me I should share with the world, they said, "Tell truth, tell all!" And that is what I have accomplished here—revealing God's truth of this strange and mystical phenomenon that only now is calling for full, unconditional public disclosure.

Kewaunee Lapseritis, BA, MS, MH
Seattle, Washington
November 2010

CHAPTER 1

CRYPTOANTHROPOLOGY: THE ANCIENT ONES

Bang…knock, knock!” was the sound heard at the cabin door late one evening by a 60-year-old divorced man living in a remote forest some fifty miles southeast of Seattle, Washington. This was in the 1990s. Upon opening the door, the man was startled to see an attractive curvaceous woman completely nude, with short brown hair (not fur) covering her entire body! There was no hair on her face whatsoever. She did not have Negroid features, wore straight long hair on her head, and had short hair that was evenly distributed on her breast and everywhere else except for her palms and bottom of her feet. The female being was around six feet in height.

The woman appeared to be pleasant and gentle, and immediately began making overtures of a sexual nature toward the gentleman. He was stunned. What little communication occurred was strictly telepathic. Soon the two people were in bed engaged in sexual intercourse. As if this was not shocking enough, after the romantic interlude, she smiled at him while climbing out of the bed, then proceeded to turn and walk straight through the *solid wall* of the cabin. Two months later there was another knock at the door, and when the man opened it, the forest being was once again standing there pointing to her belly. She said telepathically, "It took," then turned and walked away. The man never saw her again.

During the previous months, before he first met the seductive lady, he had observed what he labeled "Sasquatch" at the edge of his property on several occasions, but he had never seen one up close. Also, a week before meeting her, he had witnessed a spaceship hovering a couple of feet above a small lake near his cabin. The gentleman claimed it had a long tubular appendage submerged in the water. The following morning he noticed that the water level in the lake had dropped about a foot. He associated the UFO with the woman with whom he had the sexual encounter. It was very confusing for him, so he called a group of investigators from a local UFO organization to report the incident and to possibly receive some explanation of what had happened.

A member of the team relayed the report to me. As a veteran Sasquatch researcher, I immediately recognized the woman the man described to be an "Ancient One" and *not* a Sasquatch! I wanted to interview the man myself, but was told he had moved and the researcher did not know his new address. The investigator said the man was sincere and distressed over the incident.

There have been many strange encounters reported across the continent that have been confusing to both witnesses and researchers alike. The descriptions range from human-looking to monkey-faced, gorilla-like, and even something similar to an orangutan. Percipients worldwide on every continent except Antarctica have also reported a variety of sizes of these extraordinary anthropoids.

The commonalities are bipedalism, hairiness, a skunk-like odor, extreme adeptness at evading humans, and psychic abilities. Some are friendly; some are protectively aggressive. The beings appear to have facial diversity no different than humans. However, when a witness sees one at a distance, it is labeled (or mislabeled) Sasquatch.

From my years of study and analysis, I understand that there are four "types" of forest giants in North America, and I have seen and interacted with two kinds—the Sasquatch people and the Ancient Ones. Yet, there may be more subdivisions of them of which I am not aware. I was told by the beings that sub-races exist and are usually referred to as "tribes," because they are a people. The four Sasquatch-types in North America are:

1) the Sasquatch people, who look ape-like—with conical heads and arms down to their knees—but are actually humanoid and very mentally, spiritually, intellectually, and psychically evolved, in spite of their many animal-like qualities that allow them to better survive in a raw wilderness environment;
2) the Ancient Ones, who at a glance look like the Sasquatch, have very human faces and round heads and are also humanoid—*not* hominids or animals in any way. Their arms are not long like a Sasquatch, but are in proportion to their bodies like ours;
3) the dog-faced or baboon-like forest giants with a "snout," which are rarely reported, but are like Sasquatch beings except for facial physiognomy; and
4) the orangutan-faced one in Florida often referred to as the "Skunk Ape."

As of 2010, I have experienced the Sasquatch people and the Ancient Ones just over one thousand times since I was stunned into a quantum reality in 1979, when both ETs (Starpeople) and Sasquatch began telepathic communication with me, revealing their interdimensional existence. All of this has drastically changed my life and the nature of my reality. My focus is to report the truth as objectively as possible under extraordinary circumstances. Telepathic communication is the *key* to this entire phenomenon!

In Oregon in 1985, I began to encounter the Ancient Ones, and discovered that they too had profound psychic ability and were also affiliated with friendly Starpeople. I was told by the Ancient Ones that all of them consider themselves a "people." Anthropologists define a people as having:

1) culture—that is, social behaviorisms, a political structure, art, and products of human work and thought that, in total, help a people to survive;

2) belief or recognition in a supreme being or supernatural God;
3) language to pass on oral or written tradition;
4) the ability to bury their dead rather than leaving a corpse to nature's elements to decompose; and
5) tools to aid them in their cultural survival.

After interviewing and documenting 187 percipients (most of whom have had mental telepathy with these forest beings, some ongoing), I can state without hesitation that both Sasquatch and Ancient Ones fit the five criteria as a *people, not* animals, and certainly not monsters! Anthropologists need to accept these gentle forest giants as an uncommon ilk of true humans, living inside of animal-like bodies.

Each Sasquatch has a different level of intellect, personality, and life experiences, just like us "hairless" humans. The Sasquatch people have become my friends—no different than my Native American, Black, and Asian friends. We are all a people, regardless how each race physically looks. The Sasquatch have lots of hair, enabling survival while living a primitive existence in the wilderness through all seasons. Their animal-like characteristics also allow them to gather wild plants, growl, howl, kill, and eat animals raw.

I discovered that they can make fire, but rarely use it, because the smoke would attract unwanted attention from forest service and private airplanes ("big sky birds" to them). What animals does the reader know that can make fire? None!

Over the years, I have witnessed mind-boggling powers from the psychic Sasquatch and how they are able to tap into a quantum "frequency" that alters our physical environment. For example: five times the beings changed the weather—four times *at my request*. And there were witnesses! The very first time, in August 1979, I was alone in the deep wilderness and it freaked me out to a point that I was shaking with fear! A detailed report of the August 1979 happening is documented in Chapter 3 of my first book, *The Psychic Sasquatch and Their UFO Connection* (1998).

Another uniquely human characteristic that Sasquatch and Ancient Ones possess is their own separate languages, as they themselves told me. Both races understand basic English and some of them are very eloquent and fluent, with incredible insight and wisdom. The two races often prefer mental telepathy. They have both oral (vocal) and mental modes to communicate.

Linguist Scott Nelson, who teaches at a university, has been assiduously analyzing Sasquatch recordings that sound to him like a language! I personally have audio tapes of the Ancient Ones saying words in English and the Caddo Indian language as well. Approximately four times, I have heard two Sasquatch speaking aloud to each other—the last time was in April 2006 in Arkansas's Black Fork Mountain Wilderness about 10 o'clock one night.

All the Sasquatch people and Ancient Ones have names because they are *human*, not animals. It may surprise some people, but a few Ancient Ones can print simple sentences that invariably express great wisdom. They have difficulty with longhand, but print in a manner similar to a third-grader in elementary school when pen and paper are left in certain places in the forest. The reader must keep in perspective that there are different levels of intelligence within each of the seven races worldwide. It is unfair, if not erroneous, to place them all in one category.

In 2004 the Ancient One named Haloti left me the following note:

Some say men want
to Kill us
We are people
Not Elohim
Not Neflihim
Not evil spirit

We are First People
Seed of Adam
Not of Cain

Learn good the old ways
Time is short
Soon come last war

We all will be safe in woods

I have a pile of letters of exchanges from two of the Ancient Ones, Haloti and Pushoma, the chief. Once, I asked Pushoma, "Will I be successful conveying the truth to the public?" He replied in writing, "Finish you write book!"

These beings are nature's cosmic contradiction—a beautiful anomaly of sensitive humanity living inside an animal-like body; an anthropological wonder roaming our planet while constantly avoiding the deadly, dominant race of spoilers that we have chosen to become.

Sasquatch and the Ancient Ones are actually advanced races who see no purpose or use for technology. They have told me and others that they were deposited here eons ago by friendly alien beings from another world. In reality, they are terrestrial extraterrestrials without a green card! These hairy folks have evolved beyond us, but humanity's notion of them is confused by their excessive body "hair," which looks similar to animal fur.

For many years I have diligently listened to and read about the views of people in the circus of researchers, and have come to the conclusion that only those individuals who have interacted with the man-creatures multiple times, using mental telepathy, know the real truth of who these nature people are. Researchers who makes sideshow claims without themselves having in-depth *personal* contact via communication show themselves to be inexperienced field investigators.

During the summer of 2008, Matt Moneymaker (who has people pay him for campouts in areas where the forest beings are frequently sighted) made claims on an Internet Big Foot forum that:

> There is more than one species of undiscovered giant primate in North America. Some people may want to "believe" that...and other people may want to sell books by suggesting that (and may even include hokey sketches to illustrate different types of undiscovered primates in North America). But that is total nonsense. Those types of authors are only concerned with selling books, rather than truth or the progress of science. The BFRO (Big Foot Research Organization based in California) collectively knows far more about these animals than those authors, and we know they are wrong on that issue.

A person should not be fooled by such rhetoric, especially when money is involved. I have received four complaints from people who were angry because a group tour claiming to offer the opportunity to see or interact with Bigfoot firsthand was a farce! The Sasquatch recognize such people and cannot be bought for a price! The forest giants are not what they appear to be.

The Sasquatch people and Ancient Ones very much resent being referred to as a "booger," "beast," "animal," "ape," "monster," and even "Bigfoot" when used pejoratively! Their feet are not big to them; they are of a normal, natural size in proportion to their bodies. Many of them despise us and want nothing to do with the outer world, seeing it as hopeless because "Whites" are so horribly prejudiced.

In many respects the Sasquatch are more human than we are, because they view their surroundings in a spiritual, respectful way. I define spirituality as the more positive side of humanity where people show kindness and unconditional love for all living things, are compassionate, altruistic, forgiving, and pleasant in nature with a sense of inner peace; plus, they actually apply and live by these universal principles. The Sasquatch will know that and be attracted to such individuals.

Keep in mind, the psychic Sasquatch are the "watchers" and spiritual keepers of Mother Earth. This is not something to take lightly. It is not folklore. The Sasquatch told the Indians this and now they are saying the same thing to contactees. The romanticized monster "myth" perpetuated by mainstream researchers does not apply here. Like attracts like; opposites repel! This is a truism with these nature people. Think of one's attitude as a mirror—what one puts out, one will get back. These words are extremely important if a researcher truly wants personal success.

When these sentient, super-intelligent nature beings observe us in order to to understand our behavior, they end up avoiding our world. They know us far better than we know ourselves and feel that our egos are out of control! They see us as dangerous, out-of-control "children." We simply need to "grow up," they have said to me and others during their communications. I am the messenger here, putting their feelings in perspective so they will be better understood.

To better understand these nature people and begin to connect with them, we must stop being so anthropocentric, so racially prejudiced, for which White European-Americans are so historically infamous. Our society teaches more of material values emphasizing the accumulation of wealth, which has put our culture out of balance with nature. We have become spiritually bankrupt in many ways. Where are our compassion, our feelings, our caring and altruism?

For example, recently on the Internet "YouTube," an elderly man was shown getting hit by a car with no one coming to his aid. That could have been your father, grandfather, or you lying there. Then a freeway camera showed a dog getting hit by a car and no one stopping to help, until another dog appeared on the screen and risked its life by making an effort to pull the injured companion to the side of the road and out of danger. It took a "dumb" animal to exhibit real compassion and altruism. What does this say about us—the supposedly superior "animal"? Just what are we evolving into?

The forest beings say that our science and society will truly become modern when it focuses on our spirituality—for that is where our imbalance lies. Spirituality is having respect for all living things.

The Sasquatch elders have told me and other contactees that they are often repulsed by hairless humans, because the average person in our society has no respect for nature, animals, and the Sasquatch people as a unique type of human being. They also said that our trashing and over-exploitation of our environment shows a lack of *self*-respect. We have gravely polluted our air, water, soil, domestic animals, wildlife, and our food, which results in an increase in life-threatening diseases. So all of the chemicals, heavy metals, parasites, and human and animal waste come right back on us in a very depraved and ugly way.

But some of their people argue with the council of elders that, because our planet is dying, a few of them should reach out to certain people with hope of educating the public, at least those who are ready to listen. They want to help, but few of us are listening. These spiritually evolved Sasquatch-type races of forest people are the ultimate environmentalists. They asked me to relay this information to the human society. To paraphrase the forest beings, their message to science and researchers is: everything you "know" is wrong!

How can we show these beings respect? Certainly not by murdering one just to dissect it in a laboratory or to place it in a cage. They know this primitive approach of science and detest us for pursuing them in this manner. This is truly disrespectful. And this is the main reason a rogue Sasquatch could be so distraught that it would destroy property and even occasionally kill a hairless human who shoots a gun at it. They are fed up! However, on rare occasions some have even been "executed" by their own clan for harming another. I have been told this by two separate clans, and other contactees have told me this as well.

In 2004 in Oklahoma, the female Ancient One named Haloti said that at times their people "marry" a Sasquatch person, but that it is rare. How is this possible? Just like us humans, these races can miscegenate, because they all have 46 chromosomes, just as we do!

While conducting field work in Texas, I was told that back in the 1800s the Ancient Ones would occasionally find a black slave hopelessly lost and in want of food and shelter. They would tell him or her not to be afraid, take the escapee into their clan, teach him or her their ways, and eventually find the newcomer a partner to mate with. The giants seem to be aware that an outsider can strengthen their gene pool.

To be a successful researcher, it's vital to look for common denominators rather than racial differences, and avoid racial stereotypes when assessing these nature beings. This is no different than interracial marriages between Asian races and Blacks or Whites with Amerindians, etc. We are all interconnected in a way that science and researchers still do not understand. It is far more complex than anyone can imagine.

Amorous encounters between humans and Sasquatch-types appear to be rare but authentic, and are occasionally reported in several other countries as well. If this sounds outrageous to the reader, I make no apologies. The Europeans have folktales of a man having a sexual encounter in a forest with a "wood nymph"—apparently a misnomer. I am convinced the men met with Ancient Ones! In Russia similar tales have been documented. Yet, it is of great interest that "testimonials" exist as part of European architecture. Several countries like England, France, Spain and Germany—perhaps others—have detailed murals, bas reliefs, wall sculptures, wood engravings and statues of wild men and wild women! I immediately recognize these as Ancient Ones and not Sasquatch, because they have human faces and are not simian-looking in spite of their hair.

For example, a medieval statue of a nude hair-covered, bearded man is found in the Schnutgen Museum in Cologne, Germany. Some murals depict a giant hairy holy man who was respected for his wisdom. Another mural in the Near East clearly illustrates Alexander the Great, whose soldiers had caught a family of Ancient Ones and viewed them as evil. It depicts the man-creatures (with human faces) being thrown into a bonfire! The scene cannot be misconstrued as anything else, since the anatomy and the story are so concise. There must be close to a hundred murals, bas reliefs, and other art forms with depictions of Ancient Ones all over Europe. The Internet is an excellent source to study this art phenomenon.

Back in December 1988, a friend and I discovered two sets of tracks in the snow, one medium-sized and one child-like about 25 miles northwest of Roseburg, Oregon. I felt that we were being watched, but we saw nothing. Nine months later in September 1989, I visited the same area again while alone, and a beautiful green-eyed young woman who was an Ancient One presented herself to me. She was about five foot nine or ten, with a very shapely hourglass figure, and did not resemble the Sasquatch in Roger Patterson's film whatsoever. This sensuous woman told me that she and her son were visiting the area and had come through the "portal" from another dimension. She said that her parents were upset with her for having a child before her time, so she was spending time away from her clan. Apparently premarital sex is a problem in her society as well. With a sly look on her face, she boldly said to me, "Now I am lonely and want you to come up by the vortex to make love to me." I was shocked and did not know how to react. There was no hair on her beautiful face. She had high cheek bones with large green eyes and a flirtatious smile. I told her I had to go and help someone. She replied, "It is good to help others, but as I said, I am lonely and want your companionship."

Just two weeks before, I had a broken engagement with my fiancé and had fallen into a deep depression. I was introspecting and did not have any sexual interest for her or anyone else. In truth, at another moment in time—out of "scientific" curiosity, of course—I would have acquiesced. After all, she was a beautiful woman. There would have been more to learn about her culture and people beyond a sexual encounter. How better to get to know someone?

The place where I lived in the southwest Oregon wilderness in the 1980s was excellent Sasquatch country, and I often explored the coastal Siskiyou Mountains. In 1998 I met a 38-year-old ranch-hand who used to be a woodcutter. He told me that once, while cutting wood alone in the forest, he felt a presence of someone watching him. The air was filled with a strong aroma of flowery perfume, though no flowers were out at that time of year. Putting down his saw, he was surprised to see a lovely hair-covered woman standing behind some bushes. Without any understanding of this complex phenomenon, he described to me an Ancient One, yet used the word Sasquatch.

She had the shape of a *Homo sapiens* woman—again, unlike the Bluff Creek female Sasquatch, who was ape-like in appearance. She telepathed to the man, asking to be sexually seduced. This startled him. He became afraid and apprehensive, grabbed his gear, and quickly left the area.

In Southeast Oklahoma in October 2005, I met a hermit who lived alone in the Kiamichi Wilderness where I was conducting field work. He claimed (and I believe him) that his "girlfriend" was an Ancient One, that she visited him from time to time at his cabin at night and had slept with him on occasion. The man said she was beautiful by our standards.

So these isolated situations are happening without the knowledge of mainstream society, and this information is important anthropological data. Interracial relationships have become more commonly accepted in the last 30 or more years and the forest incidents documented here are an example of the more unusual ones. Many researchers ridicule these reports because they mistakenly believe that the Sasquatch are apes—which is speculation by those who have not carefully studied the data. It's vital to keep an open mind.

Then there is the documented and historic Zana, who was not a hominid but humanoid in my opinion. She was considered a wild woman, captured and then domesticated in an isolated village in the late 1800s in Tsarist Russia. Zana has been described in the literature as acting more animal than human. But the public has a distorted image of who and what they are when living in nature. It was reported that Zana was sexually molested by several men in the village, where she gave birth to human-looking offspring. Once again, genetically speaking, how could this be if she were not human, carrying 46 chromosomes? In fact, Zana produced several progeny, four of whom lived—two male and two female. This is part of the *history* of that region.

Some researchers theorize that Zana was a living, breathing Neanderthal, but this cannot be proven. If she was not humanoid, then how was she able to conceive a human child? Yet all of Zana's children had primarily human characteristics and were not hirsute or wild. All reached adulthood. The youngest of the four, named Khurt, died in 1954, as noted in Christopher L. Murphy's book, *Meet the Sasquatch* (2004).

I was told in July 1985 by the Ancient Ones that there are seven distinct races of their people worldwide, each at different levels of development—the Yeti (in the Himalayan region) were the least evolved with more animal characteristics, they said. These statements have been verified by other contactees who were told the same information.

Zana could have had telepathic ability, but perhap the peasants in the village were not good candidates to receive her communication. Mental telepathy is "language." I believe many (or all) of these races of hairy folks can transmit feelings and thoughts via psychic communication. They told one experiencer that it is we, the potential receivers, who are blocking telepathy with our negative thought patterns of *fear, anger, resentment,* and *aggression*!

I have read of four episodes concerning the kidnapping of an Indian maiden for the express purpose of having a new progenitor who would strengthen and add variation to the Sasquatch tribe and Ancient Ones. The Sasquatch people and Ancient Ones have told me that occasionally rogue members of their group have taken women from our society when there is a shortage in theirs. The beings insisted it was extremely rare and they do not approve of kidnapping. Some women eventually escaped to tell what happened. Most Indian women had offspring, proving without a doubt that the Sasquatch, as well as the Ancient Ones and other such tribes around the globe, are human.

Ivan T. Sanderson noted in his 1961 book, *Abominable Snowmen: Legend Comes to Life*, that all the various tribes he visited from British Columbia through Washington, Oregon, Idaho, and California told similar stories of female abductions.[1]

In the mid-1980s, in *Alaska Magazine,* an article reported that a Nome pilot and his new wife were in route to Anchorage to make arrangements for a honeymoon. Because of adverse weather conditions, the couple decided to land, camp for the evening, then fly off to their main destination in the morning. When the groom went to fetch the gear in the Cessna plane with his bride tagging behind, suddenly she screamed, and when he turned around, all he saw was a huge Sasquatch running off with his wife under one arm! The distraught husband followed as best he could, but only got a glimpse of them three separate times before giving up. The next day he flew on to Anchorage and reported the event to the Alaska State Police. A search for his wife by the authorities proved fruitless.

In Daphne Sleigh's 1990 book titled, *The People of the Harrison* (British Columbia), she states from school principal J. W. Burns' newspaper stories of the 1920s:

> About 1880, Seraphine Leon is kidnapped by a Sasquatch and held prisoner in a cave for a year. When she is about to give birth to a child, at her entreaty, she is carried home to the Douglas reserve: the baby does not survive.
>
> Also sometime during the 1890s:
>
> Charles Victor of Skwah (near Chilliwack) and Henry Napoleon discover a Sasquatch cave 5 or 6 miles from Yale, BC. Some years later, Henry Napoleon returns to the same locality and *talks* to a Sasquatch, who tells him that the home of the monster is at the top of Morris Mountain.[2]

Once again, a humanoid Sasquatch successfully mates with a human, producing a child who unfortunately dies at birth, showing that both Sasquatch and Ancient Ones are *humans*! Their excess body hair that protects them from the elements does not make them animals! Also, for a Sasquatch to "talk" to Henry Napoleon, the giant had to be human. Chances are, it was through mental telepathy, though no one was familiar with psychic phenomena back then, to properly define them. On October 13, 2004, the newspaper *Pravda* ran the following article. Instead of scientifically analyzing the data, a journalist sensationalized the story by headlining the article "Woman Tells Her Story of Being Married to Bigfoot." What the media missed was the fact that these rare miscegenations have been reported and have taken place all over the planet for centuries. Their report was merely a recent one demonstrating once again the creature's humanity. It read as follows:

> A fantastic love story has been recently unveiled in St. Petersburg. Psychiatrist Nikolai Boyarchuk said that he had copies of the text of the story from the file of a female patient. The doctor said that the story, that happened with Oksana Terietskaya, was absolutely real. He added that it would not be immoral to write about it in the press, because the woman either died or she would never return to live with humans again.
>
> The 19-year-old was "married" to the Bigfoot for almost a year. The girl lost her way in the woods one day, after she had been hurt by her boyfriend. She went to wander in the woods just because she could stay there alone with her feelings. Oksana completely ignored the fact that she had lost her way home. She sat down underneath a tree and cried, trying to get over the pain in her heart. She realized that she had gone astray when

it was too late. She came across raspberry bushes and decided to eat some berries before looking for a path home. When she was picking raspberries, she heard a strange noise nearby, as if someone was champing. When the girl moved the branches aside, she saw a big hairy creature that looked like an orangutan. The girl screamed and lost consciousness.

"I came to my senses in a cave. I could hear a stream nearby and there were rays of light coming down on me from a hole in the ceiling. Tang—that's how I called the creature afterward—was sitting opposite me. He was baring his teeth, as if he was infuriated. I realized later that it was just his smile. The hairy animal came up to me and started sniffing my clothes. Then he roared and tore my clothes to pieces. My heart was about to explode with horror, but he continued sniffing me until his nose stopped near my groin. He roared again and threw himself over me."

When Oksana woke up the next morning, she realized that she had become the prisoner of the hairy creature. When Tang went out, he would cover the entrance to the cave with a big boulder, leaving no way for the girl to escape. Tang would always bring something to eat—berries, nuts, mushrooms, eggs, or raw meat. The terrible sex with the animal became a daily torture for Oksana.

There was a spring in a corner of the cave—the water was running somewhere outside the cave. Tang strongly refused to let the girl out. The "beauty and the beast" started developing a relationship. Tang showed interest in the girl's CD player. Oksana had only one CD with her—best hits of the band Kino. When the girl carefully showed the monster how to listen to the music in the headphones, the Bigfoot was horrified. He got used to the music later, though, and even liked one of the songs on the CD. Tang was very upset when the music stopped playing because of low batteries. He would spend hours shaking the device in his hands. "I took the batteries out and gestured him that it would not work without them. The next morning Tang took one battery and left. When he returned to the cave in the evening, he brought a pack of batteries with him." The Yeti had undoubtedly broken into a little shop somewhere in the town. Oksana concluded that the cave in which she was staying was not too far from a settlement where people lived.

The Yeti's prisoner could not see how days turned into nights, and how summer turned into autumn. When Tang started stocking food for winter, Oksana figured that it was already autumn outside. She tried to explain to the beast that she was cold. Tang listened to his "wife" and left. The hairy monster turned out to be rather brighter than Oksana thought he would be: In the evening Tang brought a warm padded jacket and pants. It became known afterwards that the girl's story coincided with the story of a tractor driver, who said that a monster attacked him in the beginning of October, shook him out of his clothes and disappeared.

Oksana was happy to find a lighter in a pocket of the jacket. "I picked some dry branches and leaves from the ground and decided to make a fire. When he saw the fire, he became very excited. It seemed to me that anger and horror were tearing him apart from inside. He became very quiet; he sat down in a corner and did not make a sound. I felt sorry for him. I managed to overcome my own fear, though. I came up to Tang and stroked him on the head. He put his big arm around me and whined.

"A week later he was happy to join me near the fire. We started frying chestnuts and meat. Tang was thrilled when he tasted fried meat. I also hoped that hunters would notice the smoke coming from the hole in the ceiling of the cave, but people did not find Tang's shelter. I caught cold in the beginning of winter. Tang understood that I was ill and he tried to feed me with some roots and plants. He would hug me tight at night to make me warmer."

Oksana managed to escape from her prison only in the spring. Her relationship with Yeti had become almost perfect by that time. Tang would take her out in the mornings to see the sunshine, but he would never leave the girl alone. One day he sensed something dangerous in the air. Before leaving, he covered the entrance to the cave with the boulder as usual, but did not notice a small gap that the boulder left. It took the girl great efforts to sneak outside, but when she finally succeeded in getting out of the cave, she started running without stopping. When she saw people in the woods, she realized she was finally free.

"Her parents took Oksana to the hospital," Dr. Boyarchuk said. "The girl was mentally incompetent; all I could hear from her was that she had been married to a Bigfoot for a year. She never managed to get used to home conditions. She was afraid of going out even during the day; she was terribly afraid of the dark. In addition, Oksana could not eat normal food," the doctor said.

The girl recovered a little at a mental hospital. She told her story to her doctor and he put everything down in Oksana's file, having considered it the description of the patient's delirium. When the girl realized that nobody believed her story, she gave way to despair. She did not show any reaction to her parents when they visited her; she did not want to eat or drink.

One day Oksana started recovering very fast. She started eating, talking, and even laughing. When doctors told her that she was getting better, Oksana laughed and said that she had never been sick. She added that "he" would come to rescue her. Doctors considered such behavior the new stage of Oksana's illness and decided to isolate her in a special room. However, the girl disappeared from her ward at night in the middle of November. Someone very strong pulled steel bars out of the brick wall. Oksana's ward mates all said that a huge hairy monster had kidnapped the girl. They never managed to trace the Bigfoot because of the heavy snowfall.

Since psychiatrists are not trained in cryptoanthropology, they would not know the history of other documented cases of Sasquatch and Ancient Ones kidnapping women. It's very interesting that the young woman told the doctor that "he" knew where she was and that "he would come to rescue her." It is obvious to me that Oksana could only make such a statement because she was communicating with Tang via mental telepathy, because he did come to rescue his "mate."

The next reference to a Yeti-type taking a female human as a mate was published in *Dawn: The Review,* August 2000, about Pakistani tribal people encountering hairy folks in a region called Karakoram. The article said:

> Some claim that the wild women prey on men and wild men on local women, even that a number of local women have been abducted by the wild men over the years…

The author of the article continues by wondering where the man-creatures disappear to and offers his own theory:

> …that the "wild people" step backwards and forwards through cracks in the fabric of time, cracks they understand and we don't.

This is another bit of indigenous "folklore" in one of the most isolated mountain regions in the world, and yet, these reports match up to similar stories of other hairy-folk behavior in totally different countries.

In Vietnam there was a Sasquatch-type creature reported by several US military veterans who encountered the primates during jungle fighting in the 1960s. There were similar reports from the neighboring country of Cambodia. On January 21, 2007, the Associated Press released an article about a woman who walked out of the jungle after having been lost for 19 years, grunting while she walked bent over like a monkey. It was reported that the girl had disappeared at the age of 8 while herding buffalo. Her family had thought all this time that she had been killed by a wild animal. No one could explain where she had been or how she survived for nearly two decades, since the jungle woman didn't have the ability to speak.

After ten months of living with her parents once again, the feral woman ran away back into the jungle wilderness. It was said she could not adjust to normal day-to-day living in the small village, but appeared fearful around people. *The Daily Mail* (October 9, 2007) quotes the father as saying that his daughter "may have run away to find 'her wild man.'" When she was captured, some witnesses claimed to have seen a naked, ape-like man who managed to slip away into the undergrowth. These beings are a nature people who live by a different set of cultural rules and who know that staying hidden means staying alive.

The Sasquatch and Ancient Ones have given me and other contactees very compelling information about the origins of both them and us

An Ancient One said that humans were "seeded" here on Earth by ETs *after* they and the Sasquatch people were seeded.

If I correctly interpret what the forest giants said, the real reason there is no missing link is the fact that each "race" was genetically engineered and placed here experimentally, including us—*Homo sapiens sapiens*.

Zecharia Sitchin is now saying exactly the same thing, based on the Sumerian history in their cuneiforms. Though this subject is highly controversial and may be offensive to many Christians, our origins must be more closely examined. Now there is more and more evidence to support such a statement. I encourage people to research our origins themselves, so they can draw their own conclusions.

What is confusing in physical anthropology and human evolution is that science has made Darwinism a fundamentalist "religion"! Mainstream science makes a leap of faith from independent hominids without presenting proof of a "missing link" for each branch on the evolutionary chain. That's because it doesn't exist!

Paleoanthropolgy has erroneously labeled lower non-human primates like "Lucy," who could walk erect, as the beginning of the branch of true man. Though these primitive forms did evolve here, based on my research, they were *definitely not* the seed of "modern man."

Likewise, Gigantopithecus may well be a giant ape, a pongid that became extinct a half a million years ago, but it may not be remnants of the Sasquatch people. We may be tricking ourselves with an academic ego by trying to force a square peg into Darwin's round hole. There is a lot of speculation among scientists and lay researchers, jumping the gun to wrongly legitimize their theory that Sasquatch is a living Gigantopithecus. We just don't know. So beware of those people who label the beings as apes—animals to be shot or captured. Most researchers simply lack personal in-depth encounters.

A Sasquatch elder named Afta told interspecies communicator Kathleen Jones that their people were living on this planet during the age of the dinosaurs, confirming what I was told back in the late 1980s. In the 1930s, dinosaur and human footprints were found in limestone in and along the Paluxy River in Glen Rose, Texas. By 1970 a scientific team began a thorough investigation of the area. All the empirical data is described in the book *Valley of the Giants* by Dr. Cecil N. Daugherty (1971). Daugherty spent ten years analyzing these tracks along the Paluxy River. During that time, he recorded more than fifty human tracks in bedrock, many of them of gigantic size. The startling thing about them is that they are literally side-by-side with 161 dinosaur tracks in the exact same layer of rock!

The largest human track in good condition measures 25 inches long. There were five tracks in all at that excavation site with an average stride of 72 inches. The researchers say they estimated the "man" to be twelve feet tall when he was alive.

The question is: Are the prehistoric human footprints those of a Sasquatch or simply a hairless human giant? If it is a human giant living in the dinosaur era over 65 million years ago, where did he come from and how did he evolve to be? The Sasquatch say they made these tracks! We must give credence to anecdotal evidence, especially when it is backed up by physical evidence.

Initially, coming from a background in anthropology, I did not know what to do with this information. My struggle with this ended when I carefully read an array of scholarly books presenting similar arguments stating that each race was brought here—starting with, perhaps, *Homo Neanderthalensis* onward. The giants did not make that clear to me. Each bit of information is like a piece of an evolutionary puzzle, related to me and other contactees who were told similar information. The hairy folks cautiously share whatever they want us unenlightened humans to know.

One must always follow the evidence to where it leads. Keep in mind that the fundamentalists of academia have been "worshipping" Darwin since 1859. With the plethora of unsolved anomalies being reported on our planet, there must be other factors at play influencing the history of man and the organic evolution of the planet. Frank Zappa said: "Our mind is like a parachute, it only works when it's open!"

Keep in mind, there is recorded evidence of true giants worldwide from ancient times and they apparently were not isolated freaks or mutants, but a *race* of people. The Sasquatch also said that the hairless human giants came much later than they did. An important question is: Were the hairless human giants genetically engineered and placed here along with all the Yeti-type races, Black Africans, Asians, Indians, and Whites by advanced people from the stars? The Sasquatch say "Yes," plus there is a plethora of "folklore" that tells indigenous tribes that they were brought here in the beginning. And physical evidence in the form of hundreds of oversized skeletons has been discovered and documented on the North American continent, in Europe, as well as in the Near and Middle East.

Hundreds of giant skeletons seven to twelve feet tall were unearthed and sent to the Smithsonian Institute in the middle of the 1800s from Ohio and Illinois down to Louisiana. California, Minnesota, and other states documented these unique and little-understood findings. They were referenced in townships, county records, journals, antique diaries, and newspapers. These skeletons are a part of recorded history.

Haloti, an Ancient One, claims the Mound Builders were a branch of her race called the Adena people. She also said that the Karankawa, a fierce cannibal tribe of Indians in Texas and Louisiana, that suddenly vanished in the early 1800s, actually migrated into the Louisiana swamps to survive the infiltrating Whites. Later, some of the Indians miscegenated with the Sasquatch people and live there today, hiding out! Haloti referred to this mixed race as the "Karan," and they were said to be six-and-a-half to seven feet tall and were feared by all the surrounding Indian tribes.

I admit that Haloti's story sounds similar to the fictional movie with Tom Berenger and Barbara Hershey called "Last of the Dogmen." It's about a band of Cheyenne Indians still living their traditional way of life hiding in the deep wilderness of Montana and totally avoiding all contact with modern man. But why should Haloti create a fiction? Those swamps are unmapped and so thick no one can penetrate them. The few people who explore these regions stick to the familiar waterways.

Investigative journalist Rob Riggs in his 2001 book *In the Big Thicket: On the Trail of the Wild Man* discusses witnesses who have seen evasive, primitive-looking Indians retreat to the swamps when spotted by fisherman, second- and third-hand stories of people in the dense swamps being attacked by uncivilized-looking Indians with bows and arrows. Again, I am not trying to prove this "rumor," but the concept and possibility are intriguing. Armchair urbanites cannot grasp the still uncharted regions and vastness of geography right here in our own country. There are thousands of square miles of wilderness left throughout all of the North American continent. We should not underestimate that.

Maybe the Smithsonian Institute *has* proven that the Ancient Ones are real, but since they don't neatly fit into the Darwinian scheme and are a direct threat to their funding, perhaps scholars had the mendacity to conceal the skeletal evidence. Author Ross Hamilton seems to think so! He wrote a fascinating, fairly well documented paper titled: *The Holocaust of Giants and the Great Smithsonian Cover-Over: What Has Happened to the Skeletons?*

Through the years, I have gathered a substantial amount of information about how the Ancient Ones live. All this information was shared with me along with a few other resident-informants in that part of the south. This cryptoanthropological information comes directly from them. Over the years, I have conducted investigations in 26 states, even to the point of exploring miles of snowy mountains on horseback, camping out in ten degree temperatures.

The Texas/Oklahoma clan of Ancient Ones assigned names to their children from different cultures and languages. For example, they liked the sound of the English name Sally, the grandmother's name. Haloti is a name from their language as is Andan her father. Sumac is her mother's name. A two-year-old was Salito, which is Mexican. The five-year-old sibling is Beaver Tail (Native American), and Nashoba, nineteen, was the same—possibly Cherokee or Choctaw. Nashoba had a strange 35–40 pound coal-black pet cat named Waddo, which means "wild cat" in the Caddo Indian language once indigenous to northern Texas, Kansas, Louisiana, Oklahoma, and Arkansas before the 1800s. So these highly intelligent nature beings are adaptable linguistically.

Some of the Ancient Ones know basic English well enough to print sentences. Also, at times, they have used Ogam, an ancient form of Irish that the hairy-folks call "stick language." They said that Sally, the grandmother, learned it "when the red-haired ones came in boats up the Mississippi River in the 1800s" (the Irish?) That's all the information I could get, but what they put on paper was Ogam.

They do use tools on occasion. Some silverware left out in the

Author on a Bigfoot expedition in the Rocky Mountains of Montana, February 1981

forest with food was never returned a farmer; it was probably used to dig roots and tubers. Glass jars have been taken by them from time to time to make "sun tea" from herbs. And back in the 1980s, in Oregon, a hunter observed a Sasquatch high on a ridge carrying what appeared to be a large pot or kettle! These nature people are on rare occasions seen holding clubs. In 2005 I purchased a rugged steel-pronged harpoon to give to Andan, the head clansman, to spear fish. An informant in Texas helped by fitting it to a long shaft then leaving it as a gift from me. The report came back saying he "caught much fish."

Haloti told one person that her people live to be 125–150 years old. Every December all the clans in northern Texas and throughout Oklahoma area migrate to a giant underground shelter-cave in the wilderness for approximately three months. In 2005 there were 34 members of the tribe. In the cave where food has been stored, they have meetings, socialize, discuss what the hairless White men are doing to the land, take security measures to protect themselves from hunters, choose a mate, take care of the sick and elderly, and so on—no different than a tribe of Native Americans.

They are masters at concealing themselves, and teach their children early how to avoid us—their ever-present enemy. They know how the insensitive Whites decimated and subjugated all the Native American tribes. That is why they remain in control of their own lives. They go to great lengths to keep their society secret from the outside world.

The Sasquatch and Ancient Ones are very aware of our technology and trickery—the use of helicopters searching for them, as well as the Johnnies-come-lately with their trail-cams and other wild ideas to hopefully obtain proof. At this point, their friendship, plus communicating and sharing with them is proof enough for me. If physical scientific proof does happen, it will only be because they *gave it* freely.

The paradox is, if a person invades their space or threatens them with a gun, their protective animal side will emerge to defend themselves. Who can blame them?

On one occasion, in 2005 in Texas, three men were fishing on a lake while drinking beer. One of them spotted Beaver Tail, roughly a quarter mile from them on the opposite shore. They thought that the four-foot-tall hairy creature was a monkey, so the men headed over to try to capture it. The boat was moored in a hidden cove (I have explored that same place and found numerous creature tracks there). The group tried to surround Beaver Tail as one of the stalkers stood on a ten-foot embankment. Danger for them struck when seven-foot-tall big brother Nashoba came to the rescue by running out from behind and slapping the man across the back, sending him to the bottom, which broke his leg, ending the fiasco.

I know these details from two sources. One of my local informants overheard the story in the workplace, and then the same incident was relayed to another informant by Nashoba himself. Because I conducted field work there two years in a row, I know the exact area of the incident. Sadly, Beaver Tail was so traumatized at almost getting captured that he hid in the woods for two days and refused to come out!

News comes from other places as well. In 2008 I interviewed a man who claimed to have had an encounter with 12 to 15 Menehune in the Hawaiian Islands. He was backpacking alone into the interior of Maui when he became ill. Stopping frequently to rest because he was weak and staggering, he became aware of someone following him. Peeking from behind bushes were little hairy people about three feet tall. As he continued on the trail, some of the little guys ran ahead of him and didn't seem afraid but merely playful, yet cautious. The jungle beings did not appear malicious in any way. Finally, he was too sick to go on and went off the main trail to an isolated spot to lie down. He was horribly ill! The hiker removed his gear, lay down on the ground, and fell asleep. (The man still appeared amazed at what happened to him while sharing this unique encounter with me.) Three hours later, he awoke while it was still daylight and discovered that "someone" had elevated him onto a three-and-a-half foot high sandy bed earth structure *exactly* where he had fallen asleep! It was six feet long and three feet wide. Who made it and how? Who moved him and then returned him to the sandy "bed"? Also, there was *no* sand anywhere in the area, so where did the sand come from? Plus he was no longer sick. Was it the altruistic Menehune? They will help us if they know we will not harm them, I was told. Could the sand have been aported [seemingly transported from another dimension]? We can only speculate. Throughout recorded history many interdimensional visitors have demonstrated unearthly powers, which apparently the tiny Menehune also have as well.

In August 1999, I recall meeting a 60-year-old Tulalip Indian in Washington State, in a Laundromat. The Indian man told me that when he was 10 years old, he and two siblings were staying in a cabin with their grandmother during the winter months. One night their grandmother ushered the three children into the back bedroom without explanation. She commanded them to stay there, as she needed to do something outside. When the children heard the front door shut, they dashed out and peeked through a window. There, on a cold winter's night, stood eleven Sasquatch by the smokehouse. They were of all sizes and weights. The largest was approximately twelve feet tall. Their grandmother had unlocked the smoke house where salmon were hung and was giving them handfuls of smoked fish. Two six-foot Sasquatch were in the smoke house as the matriarch handed them what they needed, which was passed on to the larger members of the family. As the group was leaving, the grandmother caught a glimpse of the children at the window. She later told them that the Sasquatch were her friends that spoke to her "in the mind" and said the Sasquatch were starving during a difficult winter and had asked her for food. Today, there are people in different states around the nation who are feeding the Sasquatch people and have formed an amicable relationship with them.

All of this clandestine activity is happening in greater numbers in probably every state in North America. In fact, there are several experiencers who call me often, because this is happening in their lives and they ask my advice. These people treat the beings with love and respect while enjoying their conviviality. All of this social behavior demonstrates that the Sasquatch are undoubtedly a *people*!

Other Sasquatch-types, also deposited here on Earth, include the Orang-pendek of Indonesia, the Himalayan Yeti, the Yeren of China, and the smaller varieties like the Ebu Gogo or Flores Island "Hobbit," a three-foot-tall human type also from Indonesia. I am convinced the Hobbits are a hairy race of people still living and hiding in the jungle, who did not become extinct 18,000 years ago as scientists postulate. Just ask the aboriginal Indonesian people living on those clusters of islands in and around Flores. Science gives little or no credibility to native folklore. Dismissal of "tales" that native people tell investigators *is* the reason why many informants become reserved and keep valuable information to themselves.

First Nation tribes feel that Whites are naïve and do not understand different aspects of Universal Nature. Amerindians and other indigenous cultures do not need mainstream science to interpret reality for them, as they are too busy experiencing it for themselves—as I have also been. Urban man has removed him/herself from nature and learned to fear the wilds of the unknown as a place where danger lurks around every tree. Poppycock! I have had people ask me, "You are out there all alone without a gun? Aren't you afraid?" These folks are from Los Angeles, Chicago, New York, Baltimore, etc, where there is a tremendously high crime rate, shoot-outs, rapes, et al., and they ask me if *I'm* afraid of living in the peacefulness of the forest with plants and animals all around me? Our society sees things upside down!

In June 2004, Mike Wisotzke, an associate of mine from New Mexico, agreed to accompany me on an expedition to east-central Idaho. We used Missoula, Montana, as a central base for food and supplies. I had been told a specific geographic area where a race of half Ancient Ones, half Sasquatch dwelled in the Selma-Bitterroot Wilderness. Previously, Haloti said that the tribe that lived there was extremely powerful and was respected by other forest giants. This group referred to themselves as the Chuska people [pronounced Chew-ska] and were informed that we were coming into their area. They said they knew me

Mike Wisotzke resting while exploring in the Selma-Bitterroot Wilderness, Idaho, in 2004

Author with pet hawk in 1962, age 18, while living in the wilds of New Hampshire—already an experienced woodsman

as a man of truth who was helping them by educating those living in the outer world about their existence. This Idaho tribe was composed entirely of half-breed people. It was no different than visiting a small village in eastern Russia where there are Eurasian people or a mulatto population in Latin America.

Using map-dowsing, I located a vortex, an energy spot used by the Chuska people. It was at the far end of a large meadow. I brought a light folding chair to sit in, to meditate over the vortex. That day was exceptionally cold and windy, making it uncomfortable to sit for any length of time. The bitter wind was relentless while black ominous clouds darkened the landscape. It was miserable and ready to rain again. This was at 6,000 feet.

When I complained to Mike, who was just leaving to look for tracks along the brook, about the inclement weather, he chuckled, "Why don't you ask your friends to open the clouds to give you some sunshine?" He continued to laugh as he lumbered away, saying, "Well, you asked them once before at an outdoor wedding for sunshine and it worked."

As I sat shivering, I thought it was a practical idea to communicate with them about it. My first try at telepathy received a promising response. I knew they were watching us. I kept my eyes closed as I began to meditate. I requested that they open the black clouds to let the sun in to keep us warm, though I had no idea what would happen. After twenty minutes I heard someone walking up to me and, when I opened my eyes, it was Mike. "I can't believe it, Kewaunee, the clouds have opened up and the sun is shining through." I was stunned to see that a *perfectly* round 360-degree hole had opened up in the clouds, as if someone had used a gigantic circular cookie cutter. It was unnatural, but triggered our excitement. A warm sun was beating down on us in 38 degree temperatures!

Minutes later I unzipped my jacket and took off my wool cap because of the steady warmth shining through. Two hours later the opening with the perfect 360-degree circle in the clouds still remained directly above us. It was like something out of a science fiction movie. We finally folded the chairs and walked a hundred yards toward our four-wheel drive vehicle parked on a trail. When I looked back, the hole in the clouds was no more. It had transformed back to the way it was, creating once again a bleak grey landscape. Mike kept ribbing me, saying, "I'm your witness; you need to put this in your second book. Nobody is going to believe it."

I need to still emphasize that it is important not to stalk these beings, but to show love and respect without any intentions of exploiting them in any way. I left a large crystal for them as a gift, and telepathically thanked them and said, "May God bless you and your people," then departed. Deep compassion and respect is the field methodology that always works for me.

We are actually dealing with unique terrestrial extraterrestrials that have been living for centuries at the very fringe of civilization throughout North America. These people are so psychically advanced and stealthy that they have fooled science into believing that they don't exist! Now that's *really* being evolved! They are both physical *and* nonphysical—not one or the other as paranormal researchers claim. The beings live in a quantum realm that is really part of mainstream physics. A merging of humanistic anthropology and quantum mechanics will advance us toward solving this mystery.

Keep in mind, most Sasquatch don't care if we have proof of their existence or not. Some proponents say one Sasquatch must be killed and sacrificed for science to once and for all prove their existence so laws can be in place to protect them. That's ludicrous! The law means nothing to some people, because there are not enough game wardens to enforce it. Hunters are pouching eagles, our national bird, all the time to sell talons and feathers. On the other hand, the Sasquatch have taken care of themselves all this time.

Since researchers have not caught, killed, or rarely even sighted a Sasquatch after many, many years of searching, it remains obvious that these nature people possess some clever abilities that have allowed them to stay hidden from us for so long. Not only do we need to change our outmoded field methodology, but we need to shift from the way we think, to perceive living nature as an integral part of us, not something separate, frightening, and wild. Holistically and spiritually, they are us and we are them in the Great Oneness of an interconnected universe. Or as the Fourteenth Dalai Lama says: "If you see yourself in others, then who can you harm?" Researchers should experiment with these benign philosophies, integrating them into their field approach and leaving their guns, cameras, and fear at home.

Indeed, it is the human race that needs to save ourselves by:

1) ending war and serial violence;
2) stopping pollution of our air, water, soil, and food; and
3) stopping the ubiquitous bureaucratic and corporate corruption that spawns "greed."

Already there is a scarcity of food and potable water because nature can't take the abuse any more.

Some Sasquatch feel it is vital to connect with certain people in the outer world because our Mother Earth is dying! They said if we continue to destroy the planet, then we too will perish and their demise will follow. Many of them assist friendly Starpeople, because the ETs do not want to see a domino effect throwing our solar system, as well as other dimensions, into serious turmoil because of our society's negative actions. We are all interconnected—more so than modern man is aware of or understands. If modern man did understand, there would be no wars, pollution, urban violence, scarcity of food in underdeveloped countries, destruction of wildlife, and certainly not hunters stalking and harassing these giant hairy people.

It is a fact that only 4% of the universe (as we know it) has been discovered, which leaves an infinite number of possibilities—a whopping 96% to be discovered that is a part of Universal Reality.

When a Sasquatch first spoke to me in 1979, it was then that I immediately knew they were human and not animal. Also, they live in family units close to a clan that gathers seasonally into a tribe for social and political reasons. The forest giants have a culture, use tools and fire on occasion, pass on oral traditions, give homage to God—a higher supernatural being—and, amazingly, seems to know of Jesus Christ. Plus, they have an unwritten language, which they can speak verbally but more often use telepathically. A few can read and write simple printed writing, but not longhand. They also must have the same number of genes as we humans, since they can mate with us and produce a mulatto-like child in some cases.

Based on my years of interacting with the giants, they are without a doubt a higher form of human living inside hairy bodies. They are unique with a huge capacity to love and socially interact with *Homo sapiens* if we let them; if only we would stop the immature behavior of monster-hunting.

There are many readers who may not agree with me that hairy folks world-wide are a *people* whose animal-like traits enable them to survive. But they are humans nevertheless, and anthropologists and Sasquatch researchers will discover this when they expand their parameters.

When the Sasquatch speak to you telepathically, your whole world and perception of them changes forever. One soon knows that such a psychic encounter is a special gift that few people have ever experienced. In spite of all the controversy questioning the reality of a psychic Sasquatch, Universal God knows that every written word in this monograph is true. I and 187 witnesses can verify that. I stand by the truth as I have come to know it.

As Mahatma Gandhi said about truth
and presenting new ideas for change:

First they ignore you,
Then they laugh at you,
Then they fight you,
Then you win!

CHAPTER 2
THE SASQUATCH WHISPERERS

While most Bigfoot hunters (lay and academic types) are busy looking for tracks, call blasting in an attempt to lure in a Sasquatch, and hanging up trail-cams in an effort to get photographic evidence, I have been quietly interacting with the psychic Sasquatch in forest regions all over the United States. My primary concern in life is gaining knowledge so that I may help others in God's name.

To clarify my role, please understand that this book is about *them*, not about me. My focus is to convey precisely what these nature people want me to say. Without my understanding or wanting the responsibility, the Sasquatch people have asked me to be an intermediary, a voice for them to express to the outer world their feelings and concerns. This made me uncomfortable at first, because the majority of people who believe in Sasquatch wrongly think it's an animal. This I can understand. Yet, these super-intelligent beings, who are evolved way beyond us (in spite of their primitive appearance), strongly feel that Canadian and American societies need to raise their spiritual awareness, so the beings have begun contacting certain people in the "outer world," to share their message with us concerning our planet.

Those people, like myself, who are having ongoing telepathic communication with the forest giants, I call the "Sasquatch whisperers." These whisperers, as such, have developed a trust and rapport with the beings so that, at times, the Sasquatch or Ancient Ones will grant a selfless request. It has happened to me so many times that I have lost count. For me it's very rewarding and sometimes a lot of fun. And the whisperers are growing in numbers as I patiently educate people so they too can become successful contactees. This is where the future of Sasquatchery is headed, because it is the higher evolutionary step to the New Science.

Communication is everything. The Sasquatch and Starpeople are synonymous—they are one and the same, just different races working together. Their goals are to 1) save the planet from being environmentally destroyed, and 2) find Earth people who are evolved enough to work with them. They feel it's important to educate the public concerning the greatest evolutionary leap in human history by learning how to live in peace on a violent planet. Capitalistic societies are teaching their people non-spiritual values that produce mental, physical, and cultural decay that has reached a point where our Earth Mother is dying! On a familial level, from a personal aspect, who would ever want to contribute to their mother's demise?

In May 1998, while living in Tucson, Arizona, I met a woman named Joyce at a business meeting. We would get together twice a week to discuss business. She had moved into the area from Utah a month before we met and rented a large mobile home twenty miles away in cowboy country in the Sonora Desert.

One night she called me from an isolated phone booth in the desert near her home, because her cell phone was temporarily dead. As we talked and prepared for our next meeting, Joyce's voice grew into a tone of fear and anxiety. She related that two unsavory looking cowboys had begun stopping at her house asking if they could help her move in. The liquor smell on their breath and the way they gawked at her attractive features made her fearful of being raped. She politely refused their offer of help and made up an excuse to get away from them. After all, she was living completely alone in wild cactus country.

While she was on the telephone, the two cowboys had driven by three times, slowing way down while staring at her. She was alone in a vulnerable position. It was 9:30 p.m. and Joyce was frightened. The gate in her driveway was locked, but she was sure these men of questionable character would be waiting for her at her home! Then in my kitchen, while on the phone, three living Sasquatch apparitions appeared to me. They were monitoring my conversation interdimensionally, as they had been since 1979 when they first approached me. "Kewaunee, if you want us to help your friend, just ask us," they telepathed. Since I had shared with Joyce that I had written a book on Sasquatch to be released that summer, I was blunt with her about my friends wanting to help her! I suppose she could have said, "You're nuts," but her response was a resounding, "Yes!" She was desperate!

I had no idea how the Sasquatch would arrive or what they would do, but I put the phone down for a moment and "spoke" to the three figures who were patiently standing in a row. I thanked them and they replied, "Don't worry, we will help."

I warned Joyce that a brightly-lit spaceship with ETs might arrive at her place. I told her, "Don't be afraid, because they are my friends." I asked her to trust me.

After hanging up, Joyce carefully drove two miles to her home and this is what she reported to me at our next meeting. She pulled into her driveway to unlock the gate, and the light from her headlights shone on a seven to seven-and-a-half foot tall hairy man who deliberately stepped out from behind the house so she could see him; then he turned and walked back again. The being was there in his physical body less than five minutes after Joyce left the phone booth. Inside the house she felt secure, because a serene energy from his presence was "vibing" her. She said she slept wonderfully all night, the most restful sleep she had had since moving in a month before.

Interestingly, a couple of weeks later, Joyce told me that she had previously seen the cowboys almost daily checking out her place, but since the night of the Sasquatch she has not seen them anywhere. She was happy for that, yet puzzled. Since the giants have the ability to use "telepathic hypnosis" to influence people, it is possible that the Sasquatch gave the cowboys a subtle threatening "suggestion" that scared them off even though they probably had no idea where it came from. At times a Sasquatch will throw out "fear" to hunters and hikers in the forest, when they have their family nearby and don't want anyone in the area. The intruders merely have an overwhelming feeling of doom and they quickly leave the area. This is a haunting feeling that numerous witnesses have described. On the other hand, when loving feelings come to me, it's very clear and intense, sent as a "welcome." I have been told many times by them that I was most welcome in their territory and that they would watch over me during the night when I'm camping in my tent. Again, the man creatures avoid and reject those who are actively seeking to exploit them. An example of their "support" when traveling on their behalf occurred in August 2006. I was on a ten-day lecture tour with a small entourage that consisted of two field investigators who were members of MUFON (the Mutual UFO Network). They graciously taxied me around, driving me first to a UFO conference in Hooper, Colorado, several hours south of Denver. On the night that the conference started, one of the attendees saw a white Sasquatch walk past a corral on the ranch at about 5:30 in the morning. The coordinator of the conference said that, in all the years she had lived there on the ranch, no Sasquatch sightings had ever been reported—until I showed up! Also, on the second day there, three fresh Sasquatch tracks were found immediately outside the guest cabin in which I stayed.

After the conference, there was a four-day hiatus until my next speaking engagements in Pueblo, so we went camping. Three Sasquatch told me that the three of us were welcome in the San Juan Mountains in a rugged wilderness that was a part of the Continental Divide. There, 10,000 to 14,000 foot peaks loomed in the distance. A male and female Sasquatch "invited" us (literally) to camp in their area. My two companions said to ask the Sasquatch not to scare them. To this the Sasquatch couple agreed. We were *their* guests and the forest where we camped was a section of their home.

This is how it's been with me for over three decades now. I have been living the X-Files—and I don't want it to be any different! I do this to document this new area of science that is being ignored—it's even threatening to many. My job is to educate the public with the least amount of grandiloquence possible, because this is not about me—it's about contributing to society and the planet and about these remarkable human beings I call the Sasquatch people. Their sagacity is needed if we are to heal Mother Earth. It seems to be an almost impossible task, yet each person has a certain responsibility to contribute if we and other living things are to survive.

When we arrived at the trail head, there were moose and elk signs everywhere. We found an idyllic camping spot beside a mountain stream and made three trips to the car to fetch our gear. On the second trip, I received a whiff of a not-unpleasant, mild, skunk-like odor that was all too familiar to me, letting us know that the giants were indeed there. All three of us stopped on the trail because the "vibrational energy" emanating from our forest friends was so overwhelming—like stepping into a zone of high-level electricity. The male was standing in an invisible shield about 12 to 15 feet from me. Once a person knows them, this is easy to determine. We began interspecies communication. One of the guys said, "I hope they don't try to spook us!" The man-creature told me, "No, we won't. We are happy you are here and will enjoy your presence as someone we can trust."

This conversation started to bring tears to my eyes. I told him that I loved him and his people. He replied, "And we love you and will protect you during your stay." This type of conversation happens frequently whether I am in the woods in Pennsylvania, Tennessee, Florida, Oklahoma, Arkansas, Idaho, Arizona, New Mexico, or many other places throughout the Pacific Northwest and Canada. This is how the beings genuinely treat me and act. There are no monsters out there, except for misbehaving people who lack spiritual awareness.

The tents were set up and a fire pit with large rocks around it was constructed. I crawled into my tent to blow up the air mattress, as it was near dark. A loud growl-like clearing of a throat was heard by all of us. It was ten feet in back of my tent. The Sasquatch man was letting us know he was in our camp. I have heard this a couple of dozen times throughout the years. I went to sleep at sundown while my two MUFON companions sat around a fire in light-weight folding chairs quietly chatting.

The next morning one of my friends told me that a female Sasquatch had telepathed to him the night before, using his nickname. This had startled him since only his family members call him that and he had introduced himself to me by his formal name. Somehow she had psychically gotten into his deepest thoughts. They are incredible beings of great power and intellect. Ten minutes after saying his name, she snuck up behind him while remaining invisible and placed both of her hands against the back of his chair to let him know in a gentle way that she was really there.

He said he finally understood why it was so difficult for field researchers to encounter them. The guys were glad they had a Sasquatch whisperer there to help them connect and have a rare experience with the hairy folks. However, there have been a few times when the beings would not respond to my request to interact with the person accompanying me. They have their own way of thinking that is very different from American society. I have learned to respect that and not challenge the way they deal with life.

Canadian Sasquatch researcher Thomas Steenburg wrote the book *In Search of Giants* (2000), in which he has a short chapter he titled "Hoaxes and the Lunatic Fringe," which I found most entertaining. Though I have met Thom and found him to be a fine fellow, he is of the hunting variety of researcher who is aggressively looking for the ultimate physical proof. His research is based strictly on "belief" and historic data, and not personal interaction with the creatures. So he is looking for an animal. Thom writes:

> Witnesses who claim lunatic-fringe encounters call me all the time. During the mid-1980s a fellow, whose name I will not mention because I know he would want me to, would phone a couple times a week claiming to communicate through ESP with a family of Sasquatch in the mountains around Banff, Alberta. He also claimed to have spent many days living with the creatures. When I asked for the photos he must have taken of the creatures, he replied the Sasquatch had mind powers that would destroy the film in the camera, so photographing them was impossible. When I suggested he take me to see for myself, he told me that they had the ability to make themselves invisible and they only trusted him, due to his peaceful nature. It has been several years now since this fellow last called me and I hope he finally lost interest in wasting my time.[1]

One would think that, after unsuccessfully searching for the last 32 years using guns, gadgets, and call blasting, he would have figured out that the Sasquatch are an interdimensional people and not just a clever animal. But consciously ignoring the real evidence has not made the psychic Sasquatch go away! My best wishes to Thom and his partner, Rick Noll, and others who continue to enjoy their monster quest.

The person from Banff was genuinely sharing his personal encounters and friendship with the psychic Sasquatch, accurately telling it as it is! Many percipients need moral and spiritual support when experiencing these mind-boggling events, because psychic phenomena are so unfamiliar to them. This Sasquatch whisperer was innocently reaching out and sharing his encounters, but was rebuffed. Hopefully, in the future, conservative researchers can be more open to anomalous experiences.

Over the years, I have been spending more and more time on phone and in-person counseling for people who have had psychic encounters. Keep in mind that the Sasquatch are *both* physical and nonphysical beings, not one or the other. Many paranormalists believe Sasquatch is strictly nonphysical, but that is untrue. Once a person has had numerous encounters with them, experiencing the creatures on different levels, then a better understanding of them slowly emerges. Teaching people the proper Sasquatch protocol makes for a successful contact most of the time. With kindness and compassion is the proper way to interact with the hairy folks, not with violence or trickery.

One whisperer in Colorado shared her encounters with a lady co-worker who wanted to visit to experience the creatures for herself. There was a cabin on the whisperer's property at the edge of the woods that was completely furnished. Kim, the lady friend, began staying there overnight on her days off. The whisperer taught Kim how to telepath and what to say, just as I had done for her a few months before. Soon there was communication, but not before she encountered heavy, thumping footfalls, growling, and rocks being banged together, which startled her at first.

When asked if they wanted food or anything, one answered, "A salt block." The cabin guest was surprised and said, "I didn't know you liked salt." The being cheerfully retorted, "No, silly, the salt is for the deer!" Here is a case where the Sasquatch had picked up one of our idioms by telepathically listening in on people's conversations. In the winter at 9,000 feet there are elk and deer wandering through people's backyards at the edge of the forest. It was unclear if the Sasquatch wanted to help the deer with a "treat" or lure them closer for a kill. The Sasquatch's funny quip was a more human reply than the lady expected. I interviewed her over the phone, prompting her to ask more questions and giving her ideas about what to say. She was delighted and spoke very animatedly about her newfound friends.

Previously, the property owner had complained to the giants about their making so many abrupt noises at night, which scared her because it was normally a quiet house. The Sasquatch promised that from then on they would be quieter, as they did not want to upset her anymore. She thought this was a "sweet gesture" and very considerate of them. They were showing their respect by honoring her request just like any other friendly "neighbor."

John, a very close friend of mine, called me in December of 2007 to tell me that his wife was gravely ill and in the hospital. She had had three surgeries, which did not correct the situation and, a couple of times, he thought she was dying! Just when John believed his wife was improving, she would take a turn for the worse. This continued for three or four months and was very stressful for my friend. I prayed for his wife daily, yet she seemed to linger. John visited her every day, bringing her things to make her more comfortable.

He believed in the psychic Sasquatch, but his wife did not believe in the creatures and preferred not to discuss the subject. One day John asked me if a Sasquatch shaman could possibly heal her, because I had successfully sent one to a hospital twice, but only after the patient gave permission. He said her permission was unlikely. Still, in desperation, John asked his wife for permission, then called me to say she had agreed. That night in the hospital room a series of strange lights floated around his wife, which amazed her! The orbs stayed about a minute. In the morning she had improved dramatically and within a few days was released to convalesce at home. John is adamantly convinced that the Sasquatch were the primary people who healed his wife at a time when she was slipping away toward death. When I ask the beings for their help, they comply about ninety percent of the time. Why they do not in that ten percent has never been explained. I always reach out to help others whenever I can, and in dire cases I have asked the Sasquatch for help. The first time that I was aware that both Sasquatch and ETs would come to a person and dramatically heal them was when I myself had a spinal injury and could not walk. The pain was a "ten," excruciating beyond anything I had ever experienced. When I telepathed to the Starpeople, they came and healed me, and I was up walking without pain *the very next day*!

Another healing experience happened when a lady friend visited me for three days over Fourth of July 2009. From the time Tara arrived, two to three Sasquatch had come into my cabin interdimensionally to feel her energy while listening in on our conversations. This has been a normal occurrence since January 1986, when the beings said I would be monitored more closely. All this was strange to me at first, as I did not know the extent of their powers.

The second night of her visit, Tara became violently ill and began vomiting and cramping. I assisted her with a warm blanket and hot herbal tea as she lay in agony on the couch. I sat up with her into the night with little reprieve. Then Tara asked if I would summon a Sasquatch shaman to heal her. I had been tending to her needs during some intense moments, and I had not taken the time to consider asking for their help. When I looked up, there were two Sasquatch apparitions waiting for Tara's permission to assist her! She gave consent! I asked them to send a shaman or ET doctor during the night to heal her. Then I went to bed. It was 1 a.m.

The following morning I tiptoed down the steps so as not to wake Tara. Eight hours had passed. She instantly sat up on the couch with an alert smile and said, "Kewaunee, your friends came and healed me last night!"

I said, "Yes, the Sasquatch."

"No, it was an ET!" Then she related the following: As soon as I, Kewaunee, was out of the room, she felt an intense electrical energy around her body. This caused her to open her eyes. Standing over her was an alien being, in a white robe, who looked somewhat like a human. He was waving his hands over her in a Reiki-type movement. She was going to ask him a question, but he instantly put her in a peaceful slumber, giving her a restful night's sleep. Tara reported that she was one hundred percent cured. We ate a hearty breakfast of an asparagus omelet with chunks of avocado on top, multigrain toast, and coffee! So their altruism is apparent when I request help from time to time. Yet they say I am helping them and the planet by presenting the truth to those who are ready to listen.

All this contact is not about "proof" or money, fame, etc., but about *genuine* contact to enable these nature people to share their incredible wisdom and insight by raising our awareness of what life on the planet is *really* about. The Sasquatch and the friendly ETs are here as a "support group" to those who are spiritually evolving away from a way of life that is severely detrimental to all life on planet Earth. This is what many of the whisperers have been told. It is the message behind the phenomenon, whereas researchers and science mistakenly focus on the material aspects of the phenomenon itself, which is precisely why they have failed to progress over the years. The environmental sciences and spiritual growth *are* the central core of the Sasquatch as a people, which includes the Starpeople with whom they work.

On numerous occasions the giants have helped me in other ways. In July 2003 I was looking for a country home so I could move away from the city, but couldn't find a suitable place. After six weeks of looking, three Sasquatch interdimensionally appeared in my room. "Why you have not asked us to help you find a new home?" they asked. I told them I felt bothersome asking for something so trite. They responded, "It's important to find a place with peace and security, so it is not trite."

So I said, "All right, I'm asking your help. I want a cabin in the forest, on a dirt road that is a dead end; good well water; with a washer/dryer; plus, I don't want to pay more than 'x' amount per month rent." Within two days I had my place.

On the evening of the very day I found the cabin, as I was preparing a salad at the sink at my Seattle house, I gazed out the kitchen window and was shocked to see in my mind a scene of me sitting in my car with the cabin in the background, and a ten-foot Sasquatch standing, smiling, beside my car. This is visual telepathy. Then I heard a kindly voice say, "Now that you have found a place, be at peace." When I turned around the Sasquatch was standing seven feet away, slowly dematerializing. This is who they are and what they do for people they love.

In July 2010, I had a problem with the front end of my car. New tie-rods were installed, but then it steered a bit too easily and somewhat oddly. I visited the mechanic and he assured me that all it needed was for the front tires to be realigned, in spite of my insistence that something seemed loose! After making an appointment with a shop that does this kind of work, I drove the 40 miles toward Seattle using the freeway. The steering got worse so I drove at 50 mph all the way. When I arrived at my destination I was told to come back in an hour and it would be finished. But when I returned, the master mechanic had a very sober look on his face. The car was still on the lift. He said after taking off the left front wheel that the tie-rod looked weird. When he reached over and touched it, the rod fell off in his hand. Then he went on to say that he didn't know how I even made it all the way without being killed. If the tie-rod comes off, one no longer has control of the steering and one either heads into traffic or off the freeway into a tree or post! The mechanic kept shaking his head, repeating that he was amazed that

I made it to the shop. Later, when I arrived back at my cabin with the front end correctly fixed, a Sasquatch apparition interdimensionally came directly through the door in front of me, saying, "We were the ones who held the rod together using psycho-kinetic energy until you arrived at a safe place." Then he turned and shot out through the closed door. I sat there stunned, then telepathed words of thanks. They saved my life.

Once in the mid-1980s I went off a 60-foot cliff in a car while driving slowly on a snaky wilderness road. A spaceship suspended my vehicle in mid-air, turned it 90 degrees, and set it back on the dirt road again unscathed. If these Star beings have such abilities, which I can vouch for, then we are not as modern and advanced as we think we are.

I am convinced that the "skeptic" magazines displayed on store shelves are government-funded, and the articles are skewed to deliberately misinform the public as to what is real and what is not. They don't want us to be empowered. If the public knew that psi was authentic (and it is), it would empower them and change the way they perceive this work. For example, ETs are not only here, but have been here for millions of years. They have permanent bases under the oceans and in deep-water lakes, as well as inside of hollow, dormant volcanoes. This I know because both ETs and Sasquatch have told me this as fact. Many of the whisperers have discovered this along with the interdimensional aspects of the Sasquatch phenomenon. The difference between the interdimensional theory proposed by a couple of researchers years ago and my work is that this information was told to me directly by ETs and Sasquatch, and has been verified by other contactees based on what they have been told, as well as personal experiences of seeing spaceships, ETs, Sasquatch, and the Ancient Ones materialize and dematerialize! Such events have changed my life and how I perceive the world.

In the summer of 2004, I began interviewing a man named Dave in northeast Pennsylvania. He seemed very uncomfortable talking to me on the phone and was hesitant to share his encounters by mail. In October 2005, I finally met him in person in Pennsylvania. Dave appeared to be clean-cut, courteous, and reserved, and had been working at the same job for over twenty years, which showed stability. Overall, he was a very pleasant individual to socialize with. He told me that over the years he had encountered the Sasquatch people a total of 17 times. He was a Sasquatch whisperer.

A stick pile where Dave in Pennsylvania had repeated encounters with the psychic Sasquatch

More stick formations in a key Sasquatch area that experiencer Dave believes were made by the man-creatures

The first encounter frightened him, as the creature followed him in the brush parallel to the trail as he headed for his truck. As he was starting to panic, the being telepathed to Dave saying, "Don't be afraid; we would never hurt you!" He drove me to four places in the forest where he had encountered them. In three of the areas, the Sasquatch immediately telepathed to me, stating that I was welcome and that their group was there. When trekking through the autumn woods with Dave, I could "feel" them clairsentiently following us. This feeling quality becomes a part of a whisperer and there is no guess-work involved. It is pure psychic energy projecting from their minds. This is the best way I can describe it, my entire body becomes a human vibrator of energy.

Dave told me that one evening before sunset he had taken his horse out for a run and its leg had fallen in a hole. He had walked the limping horse back to the stable. He decided to call the veterinary clinic, so that in the morning someone could come over to examine the horse. When he glanced out the kitchen window at dawn the next morning, he was alarmed to see two Sasquatch walk out of the horse barn. Dave was concerned for the safety of his horse. Upon investigating the matter, he immediately noticed two things about the horse: its leg was completely healed and normal; and its mane had been braided!

On another occasion, while on his small farm, Dave was astonished to see half a Sasquatch body walk by him. It was the queerest thing he had ever seen! The hairy hips and legs were visible in the physical plane while the upper torso, hands and head were completely hidden from view. Dave explained that, based on what he observed, it appeared to be a juvenile creature roughly five-and-a-half feet tall. He surmised it was a youngster just learning to dematerialize—and he was probably right. From what I understand, some of their defensive quantum powers are natural, but must be developed to be used efficaciously.

At a later date, Dave was again perplexed to see a single giant hairy arm wrapped around a tree at a height that would make the creature seven to seven-and-a-half feet tall. Then the white-haired arm disappeared. The Sasquatch itself remained invisible behind the tree.

Experiencers need to share their weird encounters with someone they can trust. These types of reports continue to come in. I would be dishonest not to state it from the records. Objective reporting is crucial, in order to get closer to solving the mystery, no matter how far out the encounter was.

Author with Comanche, his Black Labrador who enjoyed the friendly psychic Sasquatch

Author holds a hair-like fabric from the Mount Adam Wilderness where he observed spaceships landing. Results of analysis by a top scientist said that there is no match for this fabric on Planet Earth!

Physical anthropologist Dr. Grover Krantz (deceased) often stated that a majority of the reports are unreliable and thus false information. That is nothing more than speculation and is simply not true. I always liked Grover and got along with him, but I adamantly disagree with his statement. Out of 187 witnesses I have interviewed over the years claiming psychic Sasquatch experiences, I consider two to have been fraudulent. Just because a researcher does not fully understand subjects like UFOlogy, Native Amerindian folklore, and quantum physics—all areas that relate directly to Sasquatch reports—one should not sweep the "messenger" under the rug. If a researcher doesn't believe what a person encountered, why should their lack of field experience cause them to discard what the witness is telling him? I understand the difficulty mainstream researchers have grasping this reality. More and more anomalies are being reported as witnesses finally come out of the closet. Making fun of percipients is a sign of an unprofessional researcher. We must keep an open mind when dealing with Sasquatch anomalies.

When I moved from the Sonora Desert to a ranch at the edge of the Mount Adams Wilderness in Washington State, I continued to be contacted by the Sasquatch people on four occasions over a two-and-a-half month period. My Black Lab Comanche and I lived in a full teepee and there were many frigid nights. One encounter was around three in the morning, when I got up to relieve myself. It was a full moon with clear visibility so I could see without a flashlight. I was walking back to the teepee when a voice said, "Kewaunee, I was listening to your prayers earlier this evening when you were asking for guidance. Keep working on yourself, Kewaunee, keeping working on yourself."

The Sasquatch was about 50 or 60 feet in back of the teepee. He said he was around 18 years old (our age) and was very interested in our world. He was around six-and-a-half feet tall or more. The young man told me he was sorry he couldn't stay to talk with me because he had been given the responsibility of escorting four juveniles from one part of the forest to another. When I asked where they were, he immediately projected "visual telepathy" showing me (like "live shot" broadcasting on television) the four Sasquatch standing by a barb-wire fence in back, where I often picked herbs. They ranged between five feet and six feet in height. I said, "God bless you," and he left. The way he stood during his visit was interesting. One shoulder was positioned high while the other was low so that one of his long arms came closer to the ground like a football player might do. I enjoyed his personality. He said his people knew about me and were unafraid to approach me. Also, he was comfortable because he listened to my thoughts at bedtime: I was praying for world peace, spiritual guidance, family, and friends. The fact that I am nonviolent was the magnetic attraction for him.

Two nights later, just as I was falling asleep, Comanche, my 90-pound Lab, nudged me firmly with his nose. I turned on one of the two flashlights lying beside me to see what he wanted. With his nose he kept pointing to the entrance of the teepee. He was a super-intelligent canine—almost human, as many friends have said. Each time I asked him different questions of what he wanted, Comanche did not respond. I told him to go back to his bed and go to sleep. This he did, but as soon as I got comfortable in my sleeping bag, there was Comanche's cold nose in my face! This scenario repeated itself three times and he was getting more persistent in having my attention.

Then it happened. The "rap, rap, rap" of two heavy sticks being struck together some ten feet from the teepee. Comanche, though very composed and patient, quickly lay down in front of the entrance, wagging his tail in an excited manner. Comanche had met ETs and Sasquatch before and loved them. He was very courteous with a mellow temperament. The three raps continued. It was my curious young Sasquatch friend from two nights before.

"It is you, my friend. Thank you for visiting me. I am honored," I said in my mind, for I was getting excited as well. I was still reclining in my sleeping bag and put the flashlight back on the floor beside me. The second flashlight had dead batteries and it was also by the crude bed in which I was lying.

Then, just as I closed my eyes, I heard the giant walking up to the teepee behind me. My back was very close to the canvas. Slowly I reached for my flashlight. Holding it in my right hand, I lay there and listened. Though I was in no way afraid, I was surprised when a large hand reached directly through the "solid" canvas and gently rested on my right shoulder. An electrical tingling energy encompassed my body, which I felt run down the length of my right arm. It may sound corny, but all I felt was love energy from the Sasquatch. It was his way of showing trust. Then he withdrew his arm and walked back into the forest. These beings have the ability to move through anything solid on the physical plane. Another fascinating thing was that my second flashlight with the dead batteries now shone brightly when I turned it on. The Sasquatch's energy rekindled the dead battery. This illustrates some of the Sasquatch's unusual abilities and should be noted as such. I feel so fortunate that they like me enough to approach me so often. It is they who are in control, not some researcher. Incidentally, that was the fifth time a Sasquatch had touched me that way!

I was new to the area and did not know very many people. A woman named Barbara approached me and asked if I would take her into the wilderness on a hike to possibly interact with a Sasquatch. She was peace-loving and kind, and I felt she had the qualities to connect. We took daypacks with water and a lunch. We drove to the trailhead of the Indian Heaven Wilderness and we were the only vehicle there that day. We entered a gorgeous Douglas fir forest. I stopped to gather some pipsissewa, an herb that is excellent for arthritis and stiffness of the joints. After hiking a ways, we stopped a couple of hundred feet off the trail at a hill in a pristine conifer grove. We sat and meditated to quiet the mind and get grounded. Meditation lasted about 30 minutes; then we telepathed individually, asking the Sasquatch to visit us. After that we had a picnic. We didn't hear or see any sign of a Sasquatch, so we decided to leave to hike back to the pickup truck.

As we approached the trailhead where we had parked the vehicle, we walked into a "force field" of heavy energy. It was overwhelming. People have described this powerful experience to me several times, but Barbara said she could not feel them. Two Sasquatch, a male and a female, were approximately 50 feet to the side of the pathway, blending into the trees. In fact, I am convinced they projected that we would only see trees and not them. I asked if they would go to Barbara's tent that evening and visit her. To this they agreed.

We climbed into the vehicle and Barbara said, "I'm not sure I believe you! Did they really say all that?" I assured her it was all for real. Apparently she wanted proof, or a sign that these forest beings were really there. It was 15 miles back to my teepee. As we drove down the winding road, each bend had a peculiar sign she had wished for. There were three large rocks piled obtrusively on top of one another beside the road, yet there had been none there when we drove up. I told her they had just placed them there. Barbara pooh-poohed it. In the next bend in the road a large tree was nearly blocking the road, causing her to slow down. It also had not been there before. At the third bend, there was a large raven in the middle of the road. At first the raven would not move; it reluctantly strutted to the side staring at us. It wasn't injured. Very unusual behavior for a raven. Barbara kept saying how strange it was that all these events happened one after another, all within a quarter of a mile of road. Again, I insisted it was Sasquatch using their abilities to get her attention. They are tricksters and this is how the giants operate. The last bend had bunches of flowers among a cluster of ferns on the right side of the road. There was no wind. Yet a lone isolated fern, a foot or so from all the others, was swinging *wildly* back and forth. The oddness of this unnatural swaying got her attention. Barbara was flabbergasted! It was as if an invisible force or person was playing a trick on us. And, of course, they were! Barbara stopped the car to watch the fern move back and forth. She finally said she believed! The fun-loving beings

Two orbs appeared where Sasquatch are frequently seen from the porch of author's cabin in the Cascade foothills, Washington State.

had made their point by creating a strange sequence of events along the road, which was a way of communicating their reality. That evening a Sasquatch approached Barbara's tent, moving the top of it to let her know they were there. She admitted that she was too scared to come out, yet pleased that it happened at my request. Incidentally, even adult Sasquatch enjoy teasing us with their powers. Why not? I know a lot of witty people who like to joke around as well.

A month later, I moved into a suburb of Seattle and started a regular job. A short time later, Barbara told me her elderly father was dying and he had only a few days to live. While talking to her in the house, an interdimensional Sasquatch spoke to me, saying, "I will give him a healing if her father gives us permission." So I told her what the being said. She commented that her father didn't believe in the giants and when she had shown him a copy of my book, *The Psychic Sasquatch*, he had said, "That guy's a nut!" My response was to ask him anyway since he was dying. Barbara called me later that day, saying that her dad had said cantankerously, "Yes, they can come. I don't care if I get abducted by little green men as long as someone helps me!"

That night Barbara's sister arrived from out of town to say "goodbye" to her dying father. At the hospital, she decided to sleep in a soft chair beside her dad that evening. During the night she heard her father's voice speaking and it woke her up. All she remembers is that he was sitting up looking at the corner of the room saying, "So that's what a Sasquatch looks like."

The next morning, when the doctor checked in on him, his physical condition had changed. He wasn't dying after all! The family and medical staff were baffled by his recovery! To my recollection, the elderly gent lived about a year before passing away in his mid-eighties.

There is a woman named Mary Rau who contacted me in November 2005 after reading my book. The Sasquatch people had spoken to her telepathically and she was seeking a greater understanding of the phenomenon from a spiritual standpoint. Mary's first experiences are documented in Chapter 3. Her story is an interesting one that illustrates Sasquatch behavior and how they approach humans.

The complexity of the giants' psychic ability is staggering as well as profound. I don't know what they do or how they do it, but they can communicate with the planet and all living things on it. They can also go inside your head when you are sleeping and convey messages to you.

Many, including myself, have experienced this. With Mary, a Sasquatch first "visited" her in a dream so he would not startle her in person. From May 7, 2006, she documents in her own words:

> The Sasquatch from my dream came to my new friend, Linda, as we were talking on the phone this afternoon and I was relating my dream to her. Suddenly she interrupted me and said, "Oh my God! There's a Sasquatch here! I can just make out his outline from the waist up. He says you are one of them. Is this true? Oh! He says to stop questioning and just tell you what he says."
>
> Linda continued the message for Mary,
>
> "She is one of us. We know her quite well. We love her. We are soft people. We look frightening to some, but we are soft. We do our work here and in the other dimensions. It will all work out when the timing is right. All is unfolding as it should.

Experiencer Mary Rau who had a Sasquatch speak to her telepathically four times, and has witnessed spaceships in a major Bigfoot area.

> When she is ready, we will have a beautiful experience together. No harm will come to her. Tell her we love her."

Linda was crying as she finished. She said the intense love from the Sasquatch overwhelmed her.

I have heard of similar experiences from people many times over the last three decades. All of this really has nothing to do with "belief," but is what people are experiencing over and over again. As a personal clearing house for those who want to talk to about anomalous encounters, I help alleviate their anxiety and place their experiences in a more practical perspective—we deal with the "why me?" a lot. I counsel contactees by giving them a broader perspective of the phenomenon and ideas about how best to move forward without the initial anxiety.

There are Sasquatch whisperers everywhere—not very many, but there are probably a few in every state. Most of them keep the experience to themselves. In October 2004 I was in eastern Tennessee investigating, and I met three whisperers there. They were very private people and would never go public or have their names in a book. I will refer to the one with whom I was most associated as Pam. I was invited to Pam's country home where she said 17 Sasquatch in a clan were being fed three times weekly. She knew most of the clan members by name. No matter where I go or whom I talk to, contactees with ongoing encounters say that the Sasquatches all have names. Since the creatures are a people, it makes sense that they have names like anyone else.

When I arrived at Pam's house, I laid out my sleeping bag in the upper tier of their barn. The next morning the family showed me the braided manes of two of their six llamas—the "braiding" having been done by a Sasquatch. Over the years, four different witnesses have said that the Sasquatch have braided the manes of their horses. This has also been documented in Russia regarding a similar creature known mostly as Almas.

On the second night we sat out on the back deck to see if a creature could be spotted by the "feeding station," where food conditions the giants to feel comfortable when near people. This is often called "habituation." Since I was tired from the long trip to Tennessee, I retired early—around eleven o'clock. I walked the 50–60 yards to the barn to bed down for the night. Beside my sleeping bag I had a large flashlight that projected a huge beam of light. Since I was on the second tier and the large barn door was open, I had an excellent view of the house and the people sitting on the deck. Soon I fell fast asleep.

The next morning over breakfast, Pam asked why I didn't go to sleep right away. I told her I climbed into my sleeping bag and did fall asleep right away. She asked what was I doing shining my flashlight over to the deck for a half hour when I said that I needed to get to bed. My response was that it had to be someone else—but who? There were a lot of puzzled looks on people's faces with that reply.

One of Pam's llamas, whose mane was braided by a Sasquatch

Two hours later, when Pam was standing on the back deck overlooking the field, she received telepathic communication from one of the Sasquatch. He told her that he snuck into the barn the night before and was fascinated with the big flashlight, so he decided to try it out. He said that it was fun! The Sasquatch had never seen a flashlight before, and was curious. So now the puzzle of the flashlight caper was solved. I had slept through the entire incident even though the giant was about two feet from me!

One evening Pam told me a story about something that had happened a few years before. She said she knew an elderly couple who lived in the forest on a dead end road. Their house was a shack and had several cracks in the walls through which she could see outside. One day while Pam was visiting the couple, the old woman saw "something" large walk around the side of the house. She became noticeably nervous. When Pam asked what disturbed her, the woman broke down and said that her husband had befriended a family of Sasquatch. The woman also said that she was uncomfortable every time they came around.

She went on to say that at times they had been very helpful. For example, a friend had given them an old heavy couch because they were poor. Their house was situated at the top of a very steep hill away from the dirt driveway. When two fellows came to deliver the couch, it was left at the bottom of the hill at the insistence of the old man. The guys couldn't figure out how he was ever going to get the heavy item up to the house. The elderly man kept saying some "friends" would be by later to move it. The woman told Pam that after they left, the Sasquatch brought the couch up to the house. Such amicable relationships are happening around us in rural areas, but such rare friendships are usually kept a secret. My recommendation is to be kind to everyone; be a Sasquatch whisperer by opening up your heart, not your ammo box!

This next episode was told to me with great solemnity. The head of the Sasquatch clan had told Pam that he took a gallon of antifreeze from her carport because of a major problem that had developed. One of their large males had killed a female who had two young children. This, Pam was told, was an extremely rare event with a member of their people. A council of elders discussed among themselves what "crime" had occurred and he was sentenced to death! They made the killer drink the bottle of antifreeze. For a long while the perpetrator had been hiding out in a cave, so after consuming the poison he returned to his hideaway.

Part of the way there, he had become violently ill and weak. Finally, he lay down in the cave and with remorse telepathed to Pam his last thoughts and words. She swears the following is absolutely the truth.

The dying Sasquatch said that contact had been made to his people by the United States military because our government knew the truth and importance of the Bigfoot/UFO connection. Some Sasquatch are "recruited" by a certain race of Starpeople. The Sasquatch are the ETs' "ground crew" to mine certain minerals, do the heavy work, act as sentries, spy for them, and so on.

The Sasquatch man told Pam that military personnel had lured him and two other Sasquatch into a building under the pretense of being their friend. The military used some form of electromagnetic device to block his use of telepathy and to stop him from dematerializing. Then they gave the creature food with a powerful sedative in it. Once unconscious, they injected an implant to the back of his skull in order to monitor and follow his every move. He was enraged by this deceit. Friendly ETs tried to help by using surgical means to remove the implant, but on two other occasions, a Sasquatch had died because of complications with the delicate operation.

This disheartened forest giant had become angry with his clan for accepting food from Pam and had planned to kill her husband. When he argued with a female of his group about the food, he could not contain his rage and killed her.

As he was dying, lying in a cave, the creature told Pam that he realized he was wrong and was sorry for his actions and deeply regretted it. The Sasquatch said he needed to tell Pam the reason for his mistake and that he now knew there were good people like them in the outer world. He said his life energy was fading, and this was his last statement to her and he wanted to repeat: "The military helicopter is here to retrieve my body." The implant had given away his position.

During the communication with him, Pam tried to phone a friend who knew the mountain so they could find the cave and possibly help the remorseful being. The telephone was mysteriously dead. She started to use e-mail to get help, but all the electricity suddenly went out in the house. Pam believes she was being monitored by the National Security Agency, which has the technology and ability to do that. She made me promise that I would not divulge her name and where she lives. I have been criticized by my peers for not listing precise names, places, and addresses of witnesses, so they can follow up to verify that I am telling the truth. I did not do this in my first book and will not in this one, because researchers and the media are known to harass percipients by invading their privacy. Also, experiencers have to deal with backlash in their community. With this kind of extraordinary circumstances, rules of science have to be modified to respect both the Sasquatch and their new human friends. Contactees in interspecies communication are the same as "informants" that anthropologists use to validate their field work when interacting with primitive cultures. I cannot find a reason why Pam would lie to me about the above story. Apparently, these unusual encounters occur when there is government intervention.

Every day that I was at Pam's country home, she kept telling me she wanted to take me to a special place in the mountains where both Sasquatch and the feared *chupacabra* were often seen. Chupacabra means "goat-sucker" in Spanish, because of the way it drains blood from animals. It's a phenomenon that people have begun to experience in Puerto Rico and the southern part of the United States. I had never heard of such a thing in Tennessee. Pam's close friend said that her family and friends spoke of the bizarre creatures when she was a child thirty years ago, and this particular area was prime territory for an encounter with them. She added that her uncle and a friend who camped up there several years ago were confronted by a chupacabra when its sharp claws literally slashed a large hole in the side of their tent when they were inside. It scared them so badly that they never went back up there. Pam said that the Sasquatch were repulsed by the weird creature and always drove them off. My belief is that there must be a major "portal" or vortex leading to another dimension where the chupacabras enter and exit our world. I will elaborate on this in Chapter 5.

However, Pam related to me that she had been asked by the head of the Sasquatch clan not to go up in the mountains for a few days. He would tell her when it was okay to visit that area. This request was honored. The next day a Sasquatch told her that they were having a large important gathering of tribes, and the council of elders would be there. Some were coming from western Tennessee, Kentucky, and North Carolina—seventy Sasquatch in all. He did not reveal all the subjects that would be under discussion, simply stating that it was an important secret event.

On the very last day of my week-long visit, Pam escorted me to this "special" place in the mountains. Four of us hiked down a trail in a slow, relaxed manner, soon approaching about two acres of open field. A large garbage bag full of food was dumped under a tree. I could feel the Sasquatch close by. They have a very distinct vibration that signals me that they are there. As I turned around to scan the area, I saw a six-foot-tall Sasquatch run, then leap behind a tree. Just as he was within three feet of the tree, he dematerialized! The other people in our tiny group had their backs to the tree and missed seeing the creature, who was 70 to 80 feet away. No one seemed concerned, because they all had seen a Sasquatch before. Apparently most of the seventy giants had left but there were probably six or eight watching us at that time.

Slowly we puttered along the trail back to our vehicles. I stopped to show off my Sasquatch hoot in the direction in which I had seen the Sasquatch. "Whooo, whooo, whooo," I sounded off. Everyone listened, hoping to hear a reply. Suddenly Pam burst out laughing. She looked at me and said, "The Sasquatch said to tell you, 'Kewaunee, you can do better than that!'" Then we all laughed. They have a sense of humor, no matter how sober they look. They are a people and want the world to know the truth of who they really are.

I estimate that if seventy members of the Sasquatch tribe from Tennessee, Kentucky, and North Carolina attended this important meeting, another seventy or one hundred or two hundred women, children, elderly and young adults may easily have been left at "home." That would potentially place the population on the North American continent at approximately 5,000 to 7,000 individuals; possibly 10,000; maybe 15,000 maximum. In Oklahoma and northern Texas, I am aware of 34 Ancient Ones as of 2006, based on what they told an informant. I don't know if there are more or fewer of the Ancient Ones, which are usually mistaken for Sasquatch. Also, I don't know whether the Ancient Ones live in some states and not others. But their territories overlap those of the Sasquatch and they inhabit the same forests, socially interacting with each other in a congenial way. I know nothing about the rare "baboon-face" variety.

Dr. Grover Krantz estimated that there are 2,000 Sasquatch in the Pacific Northwest, and researcher Danny Perez speculates that 100,000 Sasquatch exist in all of North America. I don't know what these researchers' sources are, to come up with their demographic appraisal, but my statistical estimate is based on communication.

Keep in mind, most researchers sit in front of a computer to search out other people's work from journalists, ranchers, hunters, and so on. What little field work is conducted is usually an interview of local people who have had an encounter or found Sasquatch tracks on their farms. Some investigators do "call blasting" of a recording made of a howling Sasquatch, in an effort to draw one near. But just what is the message that the researcher is sending?

In nearly 55 years, tons of data have been collected and documented, yet more friendly contact with the giants is needed to receive salient information from them. That is not an easy task, because the giants don't trust 99% of the researchers, so this is the role of the Sasquatch whisperers.

Traditionally, natural healers, shamans, and medicine men and women are more likely to be contacted by the psychic Sasquatch because they are compassionate care-givers who do their healing with an open heart. Also, some healers have the ability to induce an altered state of consciousness in themselves. In so doing, they reach a higher vibrational frequency that is closer to the realm of the giant forest people. All medicine people are prime candidates to be a Sasquatch whisperer if they pursue it.

Some people are very natural, kind, and altruistic, and no matter what vocation they choose, domestic and wild animals are attracted to them. But even natural Sasquatch whisperers can have mixed reactions to their encounters.

A physician in Flagstaff told me that he was hiking in a remote canyon in northern Arizona when he noticed a pathway leading up to a cave. He cautiously entered the large den only to notice an enormous man-like being hiding in the shadows. This startled him. Then a voice said in his head, "Do you have a gun? You're not going to shoot me are you?" The hiker immediately turned around and left!

An attorney in Maryland began having Sasquatch encounters on the east coast. He's not a researcher, nor did he ever pursue the subject in any way. With a lack of fear and a genuine desire to help the giants, he began to put up several "feeding stations" in areas where he found signs of them. He checks the stations two to three times weekly and brings more food. I know the details because I interviewed him at length in my home in 2007 during his three-day stay. He told me that he had experienced several paranormal events that left him confused. He observed the "levitation" of feeding sacks, a rock floating between a Sasquatch's hands, and objects moving around him in the woods that indicated that Sasquatch can become invisible at will. They had followed him home interdimensionally, which was difficult for him to understand, but they would leave things for him (as they did several times in my house); and he said they read his mind.

A relatively new researcher in this field named Joy called me from California. Actually, we first met when I was being interviewed on a radio talk show. Joy had been exploring the Sierra Nevada Mountains without any meaningful experiences. In May 2009, I invited her to visit me at my cabin in the Washington Cascades foothills, where she could camp out. The morning after she arrived, we had breakfast and then I had to make some business calls, so she went outside to brush her hair on my deck, which overlooked the forest. When she turned around, she saw, at the bottom of the steps, a four-foot-tall juvenile Sasquatch. She was amazed when it dematerialized. Then, immediately, a strange fuzzy-looking orange-colored sphere about a foot in diameter aported on top of the railing! Moments later it was gone.

Twice I have seen the beings dematerialize from a physical state—once a Sasquatch and the other time an Ancient One. This was my visitor's first encounter, and she said she will never forget it. I have learned from decades of experiences that, when a percipient has a paranormal encounter, it is so astonishing that it's etched into that person's mind forever. It was such an unimaginable event that most remember every detail and cannot be talked out of it by a skeptic.

In May 2010, I invited Joy back for a visit. Upon arriving, she reported to me that approximately 8 to 10 miles from my cabin, while driving in the country, she was overjoyed to observe a chocolate-colored Sasquatch peering at her from a side road. The Sasquatch was in a small clearing by a fallen log beside a dirt road. The sighting was so unexpected that she kept driving in somewhat of a daze.

This time Joy spent two days at my place. The morning she left, I stepped out to escort her to the car and to guide her out of the driveway. To my surprise Joy didn't seem to pay attention to my directions while backing up and wildly went straight into the heavy branches of a conifer tree! She later explained that the same chocolate-colored ten-foot-tall Sasquatch she had seen earlier was standing behind me at the edge of the forest looking at her as she backed up. This instantly discombobulated Joy and into the tree she went!

Percipients who have the "right stuff" often have repeated encounters. In the summer of 2004 an old friend named Linda called me from England. We had met years ago when I was living in Great Britain. Eventually I made arrangements for her to visit me in Washington State. Over the telephone I told Linda that the Sasquatch would be delighted to meet her when she arrived. She said that would be nice, but thought I was only joking. I wasn't! The first two days, while sitting on the porch, she couldn't figure out what the heavy "thumping" of footsteps were and twice ran inside thinking it must be a bear. This was on a sunny fall day. On the third night of her visit, Linda woke up with what she described as a "horrible sinus headache." Stuffy and feeling miserable, she sat on the couch with her head in her hands. Eventually she walked toward the kitchen door. She was shocked to see a Sasquatch standing just outside the door looking at her. Moments later, she realized that she no longer had a headache and inflamed sinuses. Linda was convinced that he healed her.

Two days later I had to fly from Seattle to the Los Angeles area for a speaking engagement. Linda stayed and rested at the cabin. The day after I left, at 4 p.m. on a sun-filled afternoon, the Sasquatch appeared on my lawn. The man-creature telepathed to Linda, sending her a long message to give to me when I returned. All of this changed her reality. In fact, when she returned to England, the Sasquatch appeared to her there once and spoke with her. When her son, who was just completing his PhD at Oxford University, came home to visit, she decided to tell him about the psychic Sasquatch. Later, she told me that he had listened with a look of wonderment and said, "But Mom, I thought it was a myth!" She replied, "I know, so did I."

There was a Sasquatch whisperer in Missouri who was asked by the giants to bring them certain herbs that they needed for a specific medical problem. They developed a friendship. An elderly man in Arkansas, who lives in the back country, was having contact and communication with a female Bigfoot. Five separate people in Texas were interacting with both the Ancient Ones and Sasquatch people. One witness told me that his interaction with them started at age 16 in Texas. He said he was swimming in a lake with an island in it and suddenly a Sasquatch swam right past him so fast that a gold medal Olympic swimmer could never have caught him. Another Texas man goes into the forest weekly to communicate with them. A high school student from Ohio meets with a Sasquatch and it has become his mentor.

Fear, confusion, secrecy, exhilaration, love, and respect are a range of feelings that most all whisperers go through. Nearly all were not looking for these beings and usually didn't even believe in them. They were taken by surprise when psychically approached. It wasn't easy for several of them. Their lives have changed because their consciousness was raised, changing the way they perceive the world around them.

A Zen Monk once said:

All life in nature has a spiritual dimension
and is sacred. When we honor other creatures
as we do ourselves, we all evolve another step.

CHAPTER 3
THE ET/UFO CONNECTION

Astronaut Gordon Cooper testified before the United Nations that UFOs/aliens are visiting the Earth. As reported in the August 15, 1976, edition of the *Los Angeles Herald Examiner*, he stated:

> Intelligent beings from other planets regularly visit our world in an effort to enter into contact with us. I have encountered various ships during my space voyagers. NASA and the American government know this and possess a great deal of evidence. Nevertheless, they remain silent in order to not alarm the people.

Actually, I am more alarmed that the government is treating the population like children by hiding these facts. Personally, I have had several visits from ETs that were of a positive nature. I have been fortunate to have seen seven different races of Starpeople, who were kind and helpful, and healed me on several occasions. With tongue in cheek, I add that the beings relieved my suffering without even asking me if I had medical insurance!

There are so many testimonials of encounters of the "fourth kind" (which is face-to-face contact) that they are literally exhausting to read. Gordon Cooper is just one of 14 astronauts who witnessed UFOs in space. Hundreds of radar traffic–control technicians, retired military intelligence people, air force and commercial pilots, armed service staff, governors, and a US president have also testified, on top of mounds of other evidence that such aerial phenomena really exists. Plus, thousands of photographs and videos have been taken by citizens all over the world. A few admittedly were found to be faked, but there are many authentic ones that we just can't explain away. In addition, these craft have been clocked at up to 17,000 mph! Check out the DVD *The Disclosure Project* hosted by Steven Greer, MD, director of the Center for the Study of Extraterrestrial Intelligence. If you are a skeptic, you won't be after viewing dozens of very credible witnesses out of 400 who have given in-depth testimonials about military projects they once worked on. Or, go to *www.ufoevidence.org*—it's overwhelming!

There are more Bigfoot creatures associated with UFOs than most researchers know or understand. In 1975, when I first read about the possibility of a Bigfoot/UFO connection, I became very annoyed. With my academic background in social science, it seemed unrealistic, if not absurd. When I first encountered the psychic Sasquatch and the ET/UFO connection four years later, it took another two years going in and out of denial to finally accept it as fact. It was not easy. Now I frequently counsel others, when they are grappling with this reality. Plus, Sasquatch people themselves have made references to their to frequent work with Starpeople during conversations with me and numerous other experiencers. The Bigfoot/UFO connection shows no sign of going away. In fact, reports are increasing.

A primary part of their agenda in connecting with humans is saving the environment and the planet. They really want us to wake up and save ourselves. They understand invisible energies and cosmic occurrences that we as an Earth culture know nothing about. The forest beings are highly evolved, but we only see (or want to see) a hairy *monster*. Because I have gotten to know some aspects of them, I see the Sasquatch with warm affection. Because of people's fear, the Sasquatch are super-selective about who they trust and will interact with. Most people are missing out on a great opportunity to grow and evolve. I have come to understand that when people change their thoughts, feelings, and attitude, they change their vibrational frequency. Too much intellectual "static" hinders a person from being more heartfelt, thus the person holds a lower vibration in which the Sasquatch and ETs are not particularly interested. One must be in one's heart space to attract them into one's life.

Our government knows all about the Bigfoot/UFO connection. They don't want to lose control of the masses, as they might if ETs told us about our true origins and how the Sasquatch are really a race of ETs that migrated here via friendly Starpeople eons ago. Many call these ideas conspiracy theories, but one has to collect all the facts.

For example, in the fall of 2003, a man wrote me from North Carolina. He told me that he was living at the edge of the Great Smokey Mountains where there is still vast wilderness in the east. I called him on the telephone and he was very congenial. Once he had read my book, he invited me out to research with him, since he sees UFOs hovering over the mountains quite regularly. Also, he said, he would feed the Sasquatch behind his house. If he missed some days and didn't leave food, they would come to his house at night and create a tremendous racket!

Then suddenly, he did not return my call when I left him a message. He had two telephone numbers and one was abruptly disconnected. The guy wrote me one final letter and I never heard from him again. The letter said that his phones were tapped, black unmarked cars followed him without trying to hide from view, and it seemed he was being harassed by one of our clandestine agencies. Finally they confronted him and told him to stop his involvement with the Sasquatch. It scared him so badly that he did.

Apparently the government is not the least bit concerned about the persistent Bigfoot hunters and those with cameras who are trying to bamboozle the Sasquatch to obtain proof, because these agencies know this type of research will never succeed. It seems to be those closest to the truth that the government agents are worried about. So there may be something more sinister here than one can imagine.

In 1980–1981 in Rome, Ohio, there were spaceships, ETs, and Sasquatch being reported by farmers in the countryside. One farmer, an ex–Vietnam veteran, climbed up on his roof at night, firing box after box of ammunition at them, but to no avail. An inexperienced researcher from the area investigated the case and got the attention of the press. Soon he too was threatened by one of our illustrious secret agencies. If Bigfoot were just a normal animal, why would the FBI and/or CIA and/or NSA be wasting so much time and money trying to squelch people who might be finding the truth?

Another man called me from New Jersey about seeing Sasquatch several times in a remote wooded area where reports involved UFOs. One evening in the winter, he and two friends claim to have observed a spaceship crash. They hiked out to the area and when they stopped to rest, one of the group members ran ahead only to return minutes later. He said there were numerous tiny ETs outside the ship and he told his friends he was going home because he was scared. All three immediately left the area. Later, he said government troops arrived and cordoned off the road until a huge truck arrived and hauled the domed UFO away. He claims he saw a canvas over the craft to conceal it.

In Thom Powell's book *The Locals*, he shares an account by a researcher and associate named Henry Franzoni:

> Franzoni tells that he befriended a man named Ed, who insisted that, for security reasons, his name was never to be associated with what he was about to share. In 1967, Ed was attached to a military intelligence unit that investigated UFOs in southern California. [Incidentally, the Disclosure Project interviewed a retired military person who was a member of a UFO crash retrieval group; Ed may have been one of the members stationed in California.] His group was ordered to a remote desert location where a spaceship had crashed and broken in two. Ed said there were four dead bodies lying around outside the craft, all were about nine feet tall and resembled what others have described as Bigfoot creatures. This witness said that each being wore a copper-colored belt with a very large buckle device with buttons on it. The Bigfoot-type occupants wore thick-soled footwear resembling a sandal. Most likely, the bodies were placed in a huge storage freezer, and one of them was thoroughly dissected from top to bottom by a team of government medical scientists. Whether these were actually Sasquatch or "drones" genetically modified by an alien race, we can only speculate. The Sasquatch and Ancient Ones I have seen were pleasant looking. Ed relates that the faces were "hideous" with pig-like noses and Mongoloid appearance.[1]

The author telepathed to the Starpeople while on a trip into the Washington Cascade Mountains asking them to show themselves. This was the second spaceship he observed that day.

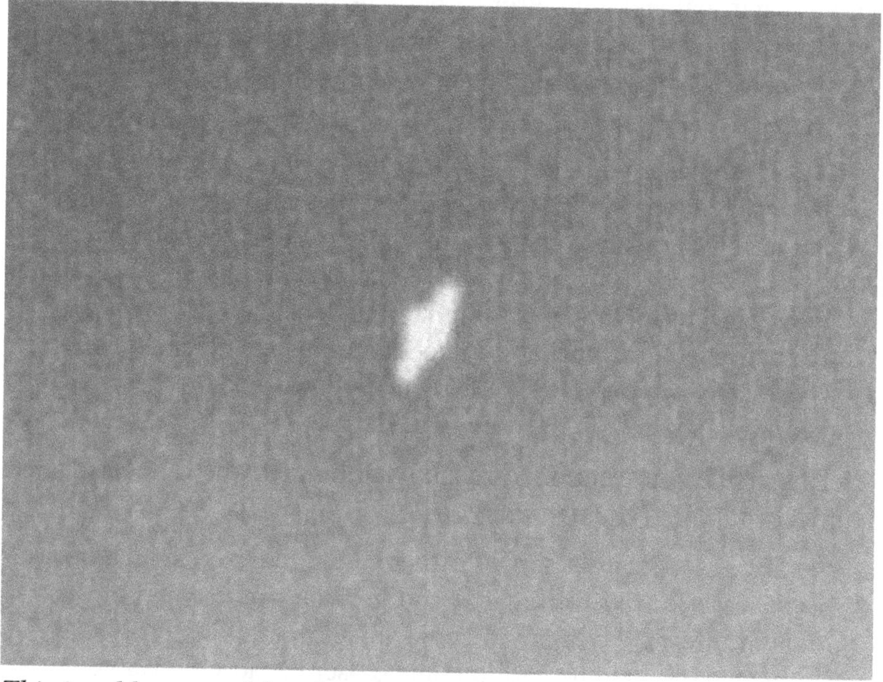

This is a blow-up of the photograph of the UFO taken on July 3, 2003.

I include such a report to illustrate the complexity of the Bigfoot enigma. Just when one feels one has a portion of it figured out, then another mystery emerges. Ed also said the corpses had a row of teeth with "stubby fangs." I interpret this statement to mean all the teeth were fang-like, which is not the norm for the Sasquatch people. One witness I interviewed intimated that the Ancient Ones had "beautiful, straight white teeth just like us."

It really bothers me that many "weird" reports are never followed up to obtain concise details from the witness; thus, *real* data like this anomaly rarely get reported. Further, it is my opinion that people who are burying, destroying, and outright ignoring anomaly reports are clearly in denial and are deceiving themselves and the public by not following up on such reports. The real "missing link" in mainstream research is the psychic Sasquatch reports, at times with a UFO connection, that have been kept secret

There was a contactee named Thomas from the Pacific Northwest who shared his personal encounters concerning a psychic Sasquatch with a researcher in 1969. It's a unique case that reveals the behavior and thinking of these giant forest people. This man was uneducated, but very spiritual and kind. From a book he had purchased, he decided to practice mental telepathy. He had heard of Bigfoot encounters in his area, which was the vast Kalmiopsis Wilderness.

> I sent off the wave of energy and thought that I would like to communicate with the Bigfoot. It took about three and a half days of sending before I reached any of them. I was working and all of a sudden, very strong, very beautiful thoughts were coming to me and from whom, I did not know for the time being—beautiful thoughts of light, of understanding, of wishing to share with me the ability to communicate telepathically. The man was equally delighted in finding someone he could communicate with in the outer world.

This gentleman went on to relate what he saw when the Sasquatch approached him for the first time.

> He was a pleasant sort of a "man," but I became frightened because he looked like an ape and here he was coming closer to me. I informed him not to come any closer, telling him I was frightened. He said he had to come closer because he wanted a better look at me. He came within 15 feet of me. I told him to

> stop or I would be too frightened to stay. He said, fine, that he was close enough. We talked over the fence for a period of what seemed to be over an hour. He told me he was the last remnant of an Ice Age civilization that was so long ago he did not even know how far back. He didn't understand time as we do.

Thomas was not a researcher and had innocently experimented with telepathy. Sharing his encounter and his thoughts is vital to understanding just who these forest giants really are.

> He told me that they were more of the peaceful type of the Bigfoot people and that there were others who were more fierce and gruesome—angry! They did not like our people taking over their land and coming into their forest community. They did not wish to be bothered by the "White man," as they called us.
>
> They wish to live their lives in peace and harmony without being bothered by other people. They seem to have an intelligence and are closer to nature than we are. I have also been told by the Bigfoot that even space beings come and visit them, where they are afraid of us. I find it very becoming that they are peaceful, loving people who are not as advanced as us in some areas, but are advanced beyond us in others—such as mental telepathy, for example.
>
> The Bigfoot said that it is not easy for us to find them because of their ability to read our minds. If you were to go out looking for a Sasquatch, you would not be likely to find one. Because, if you walked along with the idea of finding one, you would not be able to find them because it would have been forewarned by your thoughts.

When he asked the Sasquatch about our human origins, this is what Thomas said was shared with him.

> Our people, our race of people came much later than the Bigfoot. I asked him if some of them had branched off and started a race of their own. He said that some did and some didn't...*some were brought here*. Apparently, he meant that space people brought their people here to live.[2]

So we are hearing the same theme over and over again about a UFO connection, even back in the 1960s when no one considered such an association.

In the early 1970s in Pennsylvania, Bigfoot and UFOs were in the news. Three witnesses saw two Sasquatches approach a landed spaceship. This incident has been highly publicized in many books pertaining to this phenomenon. UFO researcher Stan Gordon has written an excellent detailed account from his files—a complete book on the strangeness of that period called, *Silent Invasion: The Pennsylvania UFO-Bigfoot Casebook* (2010). This superbly documented monograph is what conservative skeptics need to read, because the state police were involved. On page 145, Gordon quotes from his files a case about three women in a car:

> [T]hey were traveling on a country road near Penn when they observed a large UFO which was on the ground. The object was described as metallic and rectangular in shape. The witnesses slowed down to observe the object. While they looked on, a door-like structure suddenly appeared, along with steps that led from the doorway.
>
> Then the most unusual aspect of the encounter reportedly occurred. The women said that exiting the doorway were two very tall, hairy, Bigfoot-like creatures that ran down the steps and continued into the nearby woods. That was the last they saw of the creatures and UFO, as the women sped away from the area.

A lot of the data in the book is being released for the very first time, and gives the reader tremendous insight into Sasquatch behavior. It is a very valuable piece of work.

Another well documented case in Gordon's book took place on the night of October 25, 1973, in Uniontown, Pennsylvania. Numerous farm-folk observed a spaceship that eventually landed in a field. Three inquisitive witnesses went to the area to investigate the craft. Upon arrival the people observed, in disbelief, the glowing dome-shaped craft illuminating the entire area of the farm. The object was said to be approximately 100 feet in diameter.

Then, as if such an awesome sight was not enough, the three percipients were stunned to see two very tall, hairy bipeds walking along the fence about 75 feet away. The creatures had glowing green eyes. One of the observers had a rifle, and in a panic, fired two tracer bullets over the heads of the Sasquatch. Instantly, the spaceship dematerialized, leaving a strange glow of soft light where it had been. Also, the two giants turned and walked back into the woods. The trio returned to the farmhouse and called the Pennsylvania State Police.

When the officer was brought to the field of the encounter, the spherical glow from the UFO was still there. It was noted that the remaining light was strong enough to read a newspaper by it. After a brief investigation by the state trooper, they were walking near the fence-line to leave when heavy footfalls were heard coming toward them. When the officer shined a flashlight in that direction, an eight-foot-tall Sasquatch was standing ten feet from the witnesses on the other side of the fence. This scared the farmer, so he fired his rifle directly into the creature! It merely reached out toward them, whacking the fence, and then proceeded to walk back into the forest. This case is perfectly documented by credible people, leaving no doubt of a Bigfoot/UFO connection.[3]

Great Falls, Montana, was also a good place to conduct field work in August of 1976. Local folks were reporting numerous sightings of UFOs. When I arrived, I went immediately to the Cascade County Sheriff's Department, where the national press cluttered the front of the building, making it more of a media circus. I asked at the Sheriff's Department for a copy of their files on all the Bigfoot reports their department had written up. The sheriff told me to come back in one hour and the information would be available.

On the reports, names of witnesses had been crossed out to maintain their privacy. One report written up by a deputy said that a husband and wife had been driving in the country when they spotted two large hairy primates walking in a field near the road. The man had stopped the car to get a better look, when he observed a round, silver object hovering 52 feet off the ground at the far end of the field. The two creatures were walking directly towards the ship. In his excitement, the witness climbed the barb-wired fence and started running after the Sasquatch to get a better look. The largest Sasquatch turned around and started walking toward the man. This frightened him and he immediately ran back to the safety of the car. The two beings continued on their way as the couple drove off.

My questions: How did the witness estimate that the UFO was exactly 52 feet off the ground unless he previously had air force or army artillery training? Why not 50 or 60 feet? Also, why didn't the couple stay to see the giants board the saucer? Perhaps they were so shocked by the sighting and intimidated by the Sasquatch that they decided it best to leave the area. But the importance of all this is that it happened and was documented by the local authority as having two reliable witnesses. This report is still in my files.

In addition, that very week, a book about the events that happened that year was released by a local author—none other than Captain Keith Wolverton of the Cascade County Sheriff's Department. The title was *Mystery Stalks the Prairie* (1976).[4]

In Snohomish, Washington, just before Easter in 1977, Steve Bismarck reported that while clearing brush in the forest in back of his home, he observed a Sasquatch being deposited there by a spaceship. Later he saw ETs at close range.[5]

The entire story is described in great detail by investigative journalist and Emmy Award winner Linda Moulton Howe in her book *Glimpses of Other Realities, Volume II* (1998). She is not a Bigfoot researcher, but has conducted extensive investigations into cattle mutilations and the ET/UFO phenomenon. During her inquiries, she discovered—independently of any Sasquatch researchers—an undeniable Bigfoot/ET/UFO connection.

There were more cases privately published by Dr. Paul G. Johnson and anthropologist Joan Jeffers in a 94-page booklet titled *The Pennsylvania Bigfoot* (1980). Their data also has several accounts mentioning red glowing eyes, one Sasquatch holding a luminous sphere, and a few reports by farmers seeing UFOs on their property and then a giant hairy primate a day or two later.[6]

UFO researcher Peter Guttilla published *The Bigfoot Files* (2005) and has cases that leave no question of a Bigfoot/UFO connection. Guttilla has documented an October 1993 account:

> Near Atascadero, California, Western Bigfoot Society, Alien Report Volume 2, #5, an unidentified witness, Nick M. was watching TV when he saw a flash of light in the window through the curtains. Jumping up to get a better look, he saw a glowing red light on the hillside in the trees. From his back door he could see a UFO on the ground in the trees. A door of the craft opened and two Bigfoot-like creatures came out and apparently took some soil samples. The creatures went back into the object, which then shot up into the sky and vanished. The witness checked the area but found nothing.[7]

Also, UFO researcher Tom Dongo and rancher Linda Bradshaw teamed up to write a book called *Merging Dimensions* (1995). It's about Bigfoot, ETs, and UFOs with multiple encounters of the twilight zone kind, including those on the Bradshaw ranch just outside of Sedona, Arizona, with a female Sasquatch who repeatedly visited them and was always a welcome guest. Bigfoot researchers would benefit from reading this book, because it mirrors many encounters I have documented in my book and shows how Linda Bradshaw and her family dealt with these fascinating anomalies.[8]

Teluke: A Bigfoot Account (2008) is a wonderful series of profound events that were experienced by the author White Song Eagle of Indiana. She befriended Teluke, the white Sasquatch, interacting with him on a farm for over a year. I first met White Song in 1988, and she was kind enough to share her psychic encounters with me. Many of her experiences parallel my own and those of more than 100 other contactees I have documented, so I know White Song's spiritual saga is true. It also includes personal meetings with Teluke.[9]

More and more witnesses are reporting these encounters. There are geographic areas all over the world where paranormal events occur on a fairly regular basis. One such place is a remote ranch in Utah, which is the subject of Colin Kelleher and George Knapp's book *Hunt for the Skinwalker* (2005). It is a place where UFOs, mysterious giant black wolves, and Sasquatch were often encountered.[10] I was appalled that researchers at the ranch went gunning for any living anomaly that they could find. And, as usual, with their violent approach, they never solved the mystery. They didn't know what they were doing and the basis for their actions was *fear.*

All those old veteran researchers who have now passed on, left this world without even getting a fleeting glimpse of a Sasquatch and with no idea that they were dealing with an evolved people. I sincerely like John Green and he deserves kudos for his years of patient and hard work. During the times we have talked, we have never argued. But a few years ago over the telephone, when I said to John that the psychic Sasquatch with a UFO connection is authentic, that I and now close to 200 witnesses had experienced a genuine scientific anomaly, he replied, "Maybe the ones down there are, but those up here are normal animals."

One night in 1989, while I was living in a wilderness cabin in the Oregon Cascades, ETs came to me. We spoke briefly. White Song was on my mind. So I asked the Starbeing if he would go and visit her. He said he would. At the time, I wasn't sure if he was placating me or would really go the 2,500 miles to Indiana where she lived. The very next day, I received a phone call from White Song, saying she had been visited by the Starpeople. She related that she had been taken aboard their ship and shown a hybrid child who looked like an 8- or 10-year-old in size. The being told White Song that it was hers from when they had artificially inseminated her years before. The spaceship was in a farmer's field. That night the farmer hadn't been able to sleep, so he had gotten up and walked around. From his vantage point he witnessed the landed craft. The incident and witness's account have been documented by the Mutual UFO Network. White Song is an excellent writer, who really understands the underlying intricacies of the Bigfoot/UFO connection. Her book is a must-read for any serious researcher in these two fields of study. Sadly, she passed away on December 13, 2010.

Without initially realizing it, I am being used in a positive way as a facilitator and go-between, to introduce nonviolent people to this phenomenon. The Sasquatch want to connect with people who have the "right stuff"—at least, gifted individuals, however they are defining them. I continue to educate the public about the psychic Sasquatch and their UFO connection. In June 2005, I was off to Hawaii to lecture, as well as to interact with the Sasquatch people on two of the islands. Elin, the Sasquatch who visits me at my cabin in Washington State, has made introductions for me when I travel, and messages the beings give to a friend in Oregon are passed along to me. One does not "interrogate" a Sasquatch or inundate them with demanding questions. They are the ones in control, and answer the questions as they choose.

There are twelve Sasquatch on Kauai and six on Maui—but it is safe to share this information publicly, because in those rugged canyons and jungles, a Bigfoot hunter would only attract mosquitoes. At a Kauai hotel, two Sasquatch astral-projected into my room on two consecutive mornings to greet me. This may sound strange, but I am used to their interdimensional entry. They were waiting to guide me to a specific place in the jungle where they would approach me in physical form.

Eventually I found a spot in the jungle to camp. They indicated that they would come to me. I put out smoked fish and a large chocolate bar, which I left for one week, but they never took them. On the first night of camping, one of the Sasquatch reached into my companion's tent and gently rocked the person's shoulder three times, which was unnerving. There was no doubt after that, that I had found the right place. The second night I heard someone walking up to my tent with an occasional branch cracking. "If it's you, dear friends, please come over and touch my tent," I said telepathically. I listened intently for about a minute or more. Nothing. Finally, I thought I had just been hearing things and closed my eyes in preparation for sleep. Whack, whack, whack—the Sasquatch quickly struck the top of my tent directly over my head where I was lying. I nearly jumped out of my sleeping bag! He or she startled me. But I had asked for it, so I ended up chuckling to myself and pleased that contact was made at long last, even if it was a simple whack. They often behave this way. They are always very coy and reserved. Their presence is very intimidating, of which they are aware. If they were to simply walk up to a person during daylight hours, that person could easily have a heart attack or go into hysteria! Their immense size, bulk, and facial features would shock the body/mind!

During my stay on Kauai, I had horrible lower back pain from an old injury that had turned into a ruptured disk. This confined me to a small stretch of beach. I didn't swim to avoid exacerbating my condition. As the days passed, I became frustrated because of my physical limitations at that time. So one night, I decided to telepath to the Starpeople for a healing, as I had done several times before. Because of the severity of my medical condition, I thought that the ETs might have more advanced technology to deal with the crippling pain!

After telepathing, the last thing I remember was starting to make the request to them a second time while lying on my back in a sleeping bag. When I awoke, it was still night and I found myself outside my sleeping bag with my face against the side of the tent. I never toss in my sleep or get out of my bag. Then I realized that I had no more intense back pain! I had specifically asked to be beamed up and for them not to erase my mind of the experience, but I don't remember a thing. Checking my travel clock, I discovered I had been gone exactly one hour. Neat!

The next day was a whole different situation. I immediately went for a hike because my back was finally normal. I climbed the rock steps up to 3,000 feet at the Napali Coast, managing to hike three and a half miles round trip. It was exhilarating and the view was spectacular. After returning to the beach to cool down, I swam for nearly an hour. I never take my relationship with the Sasquatch and Starpeople for granted, but thank them over and over again for their love and friendship. The healing was near miraculous, and I was able to better enjoy the rest of my month-long stay in the Hawaiian Islands.

The Sasquatch on Kauai were shy and were not used to being so close to people, as they live in the rugged interior. There are portals on both Kauai and Maui that they use from time to time, leading to another world. I don't know what's on the other side (more on the vortex phenomenon in Chapter 5). It is important to thoroughly understand that, though the beings are benevolent, they are fiercely protective of the private areas where they live and have family. Intruders with the wrong intentions, beware!

In May of 2004, I met a medical student named Sarah who was interested in herbology and who asked me to share my knowledge in that field with her. We interacted socially and academically about twice a week. After one month, Sarah called me to say that she just had an EKG and the finding was a serious heart problem. Though she had conveyed to me that she wasn't feeling well, Sarah still planned to fly to Phoenix, Arizona to assist two physicians who were going to conduct a three-day seminar. I was worried.

Now, Sarah knew about my research and that I had published a book on Sasquatch. She had a cursory interest in the subject, but we

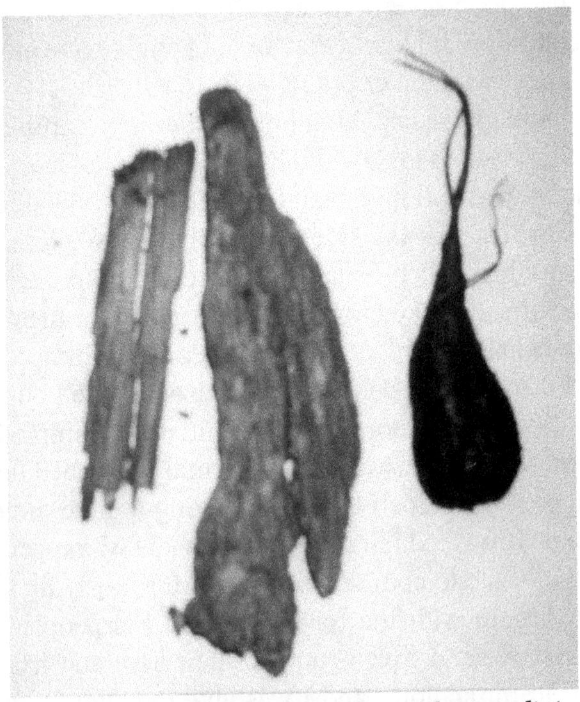

Three different herbs indigenous to Texas that medicine woman Haloti, an Ancient One, gave to the author to use for his ruptured disk. Because many gifts had been left for her, Haloti wanted to help heal the author's troublesome back problem.

had mostly discussed natural healing modalities from a professional standpoint. I had enough experience in the health care field to know that Sarah was in trouble and could easily have a heart attack. I wanted to help her, but I wasn't sure how. Then I thought of the Sasquatch and ETs who had healed me in the past at my request. Usually, when I contact the Sasquatch people to ask them to go to someone, I need to get a person's permission first. I telepathed to them asking for one of their shamans to go to Sarah, emphasizing that it was urgent! I explained the situation, hoping they understood my feelings in the matter. They did.

Before she left, Sarah had said that when she got back, she would call me Tuesday morning to confirm our meeting for that evening. I was to deliver an impromptu talk on certain medicinal herbs that she needed to know about. On Monday afternoon, I drove into town, picked up my mail, and headed home. Upon arriving, Sarah was waiting for me in the driveway. She had a sober look on her face, ushered me into the house, and this is what she told me.

On the last day of the seminar, her job was finished at 10 p.m. Sarah and one of the doctors from Seattle decided to go for a ride in the desert after three days of intense work. They took a back road, parked, then went for a stroll among the giant saguaro cactus. Eventually the two found a spot to sit on the ground and talk. In Arizona's Sonora Desert the stars are so clear that they seem to leap out at you at night.

As they were sitting, the doctor noticed "two people" walking through the desert at a distance. According to Sarah, he commented: "What would two people be doing walking around out here at this hour of the night?" A few minutes later he said, "Gee, they're walking this way." The next words the doctor said were, "Oh my God, they're Sasquatches!" Apparently his fears overtook him and he kept repeating, "Don't look at them," covering his eyes. But by then the two giant figures were standing in front of them. Sarah blurted, "It's okay; I bet Kewaunee sent them."

The tallest being was twelve feet tall. She said there was a 20–25 foot tall saguaro behind him that helped to estimate its height. The smaller one was approximately eight feet in size. She believed the big guy was a shaman and the other was a guard. The shaman knelt down and said telepathically, "Kewaunee sent me to heal you. Do you give me permission?" Sarah agreed and stretched out on the ground. To her amazement he took his hand, went right through the solid wall of her chest and gently wrapped his fingers around her heart. She told me she could feel him moving his fingers around inside her torso. There was no pain, just a vibrating sensation.

The procedure was similar to "psychic surgery," except there was no blood coming out. Personally, I have had psychic surgery five times from an authentic healer wearing a short-sleeved frock. My eyes, at times, were one foot away from the healer's hands when she opened me up and the blood came oozing out, not from her hands or anywhere else. And the healing effects on my body were profound! Apparently this giant being was doing something similar but more complex.

To complicate matters, as the physician-friend sat freaking out, an evil-looking ET suddenly appeared about 30–40 feet from them, she reported. The Sasquatch told Sarah that he must quickly finish and return to the spaceship hovering above them because he was in danger from this malevolent ET. He withdrew his hand, stood up beside the sentry who was constantly scanning the surroundings, and both of them dematerialized in front of Sarah and the doctor—just like on Star Trek!

Finally, the two of them became aware that they were just sitting there bewildered and speechless in the quietude of the desert night. They had lost all sense of time. The physician became aware that too much time had passed, and he wondered why. To this Sarah concurred. She was aware of "something" in her abdomen and immediately became upset! An ET voice said, "Do not be concerned or disturb the implant we just put in you; it is the same kind that is in Kewaunee. You have been aboard our ship all this time, which is hovering way above you. Now we can respond quicker to you if a medical crisis occurs. Now go in peace." I could see how cathartic it was for Sarah to tell me the details of her encounter. Her facial expressions were strained, as she was still in awe of what happened to her. Later I interviewed the doctor and he independently concurred with the narrative given to me by Sarah.

While visiting the scenic northeast corner of Pennsylvania in October 2005, I had the privilege of meeting Priscilla, a factory worker who lived in the rolling hills deep in the country. She worked the night shift and drove both to and from work in the dark during the winter months. The day we met, I interviewed her for three and a half hours. Priscilla was suspicious of me, so it took over an hour before she opened up. She thought she was the only person to have these "weird" experiences.

I conduct "soft" informal interviewing, where I socialize, am witty at times, and let the interviewee simply share at his or her level of comfort as I ask questions without formally interrogating the person. Being respectful is important. Priscilla said she kept a journal of her encounters and telepathic conversations with both Sasquatch and ETs. She seemed surprised that I had written a book combining the two anomalies. On her way to work early one morning, she had discovered just how real the psychic Sasquatch and ET/UFO connection is. The following is taken directly from my notes of that day.

It was July 2004, at 5:55 a.m., as Priscilla remembers looking at her watch. The weather was dry and warm. Her Ford pickup truck suddenly went airborne at 30 mph while rounding a curve at the top of a hill. Just seconds before, she had observed a spaceship hovering 5 to 6 feet above the road. Also on the road she was shocked to see 5 or 6 Sasquatch and approximately 14 ETs! At that moment the truck spun out of control and hit a tree (I later found and examined the damaged bark on the tree to help substantiate her claim).

Priscilla said she was stunned and felt both pain and blood on her forehead! Immediately her door was opened by a kindly 6½-foot-tall Sasquatch with a human face (an Ancient One?) As she moved about and was assisted out of the truck, the pain on her face intensified. The man-creature gently placed its hand on her forehead and the discomfort and bleeding abruptly stopped. She reportedly had no fear, but was more amazed at what was taking place.

The contactee continued by telling me that she was "levitated" to the ship, then laid down inside on a "table" in a dark room, where a light shone down on her. That was the last thing Priscilla remembered.

Her next memory was of knocking at the door of a country home for help, not knowing how she had gotten there. When she arrived at her house, her truck mysteriously was in the driveway. She wondered how *it* had gotten there. While taking a shower, she examined her body and discovered several puncture marks on each side of her abdomen as well as on the inner side of both thighs. Plus, her chronic blood disease (excess iron) was inexplicably gone after the encounter, as noted by her physician!

After this first encounter, she claims to have had ongoing telepathic communication two to three times a week from the Starpeople and, at times, from the Sasquatch as well. The ETs told Priscilla her life's purpose and what would happen in the near future. She declined my request to read her journal.

Priscilla was still having psychological repercussions from the experience that had happened a year and a half before. She spoke of her Bigfoot/ET/UFO encounter, saying, "At the time, this was all so new to me and I was literally going through a breakdown and believed I was going crazy." Priscilla had struggled with all this and tried talking with her partner to whom she was close, but he consequently left her, not believing what had happened to her. Even her children became afraid of her. This hurt her deeply, she said. She just didn't know who to trust.

She managed to get counseling and, later, was helped by an Indian shaman who helped her put the encounter in a healthier perspective. She kept visiting the shaman, who knew all about the "sky-people" and the hairy ones. Now she is beginning to heal. In a letter to me, Priscilla wrote that she finds peace of mind walking into the woods by her house to meditate. One day while using her Native American drum with her eyes closed, she

> …felt this strong tingling like I do when they are present, and a smell. I slowly opened my eyes and about 20–30 feet away on a big rock were three Sasquatch and they seemed very soothed by the drumming, almost like it put them in a trance. I watched them until the music of the drum stopped and then they put their hands up, palms out, arms extended straight in front of them and turned and disappeared. I was not afraid. I was slightly enamored by them and happy to experience that moment.

Back in January 1997, when I was living in Tucson, I met a couple who had recently moved to the area and were living on the outskirts of town. My first book had not been published yet, but during our meeting I shared the fact that I was busy writing a monograph. The woman said she had never heard of a Bigfoot/UFO connection and knew nothing about the subject. We met a second time and I was slowly getting to know Cheryl and her husband.

Then one day Cheryl called me excitedly saying she had an incredible encounter. She told me she had gone outside one evening to get some fresh air. Looking up, she had seen a UFO going back and forth in the night sky. She was thrilled. After it was gone she began to feel a strange "energy" around her. I quote her here from a diary she shared:

Yowie expert Tony Healy from Autralia visiting the author at his cabin in Washington State

A male and female, 2-foot tall nature beings with small tails encountered by Dick Robinson of Utah

> I began to feel some fear and the next thing I heard was, "We hear you are a friend of Kewaunee's. Don't be afraid; just let your fear fade away. Feel the peace that is pouring over you." I was really struggling with the fear, because I saw before me three very large masculine Bigfoot creatures. One was larger than the other two. The largest one was eight feet tall, and the other two were around seven feet tall. It took all my power to concentrate on their simple words, and then I thought to myself, "this must be what they consider telepathic communication to be." To say I was stunned is putting it mildly, but I also began to feel my fear fade away. It felt as if I had stuck my finger into a light socket. My whole body was vibrating with their energy, but I felt safe.

Cheryl told me that they spoke to her briefly and were gone as fast as they had come.

Internationally, it's interesting that Russian hominologists (those researching northern Asia's Bigfoot types) are divided about the connection of ETs and the Almas. I have not heard of that in China, but possibly the data is being repressed; we shall see in time. Australia has Yowie/UFO accounts, but these are very rarely documented.

In June of 1988 in *Magical Blend Magazine*, there was an intriguing write-up by author Eugenia Macer-Story. It was an interview with an Australian Aboriginal medicine woman, Lorraine Maffi Williams, in Woodstock, New York. The indigenous people "down under" have an ethno-history that goes back several millennia, farther back than most anthropologists dare admit. Besides Williams' sharing about the importance of the Dreamtime, the spirit realm, cultural beliefs, and their responsibilities within the social structure of her tribes, she talked about the tribal origins and *planetary helpers*, especially when the Earth becomes out of balance from the abusive behavior of dominant technocracies. She related:

> In ancient times not only the seven spirit brothers came to Earth, but two other different races there were, like refugees, also from a disaster planet. In this country I believe they call these beings Bigfoot. In Europe, the Yeti. In my country the White people call them the Yowie. But we call them the Dulligan. They were like liaison people from the planetary helpers [Starpeople] to us human beings. In my country in the hills the young men and older men used to go and receive teachings from the Dulligan

> people. The Dulligan were also hosts for other planetary helpers. When other planetary helpers come down to Earth, they did the job in secret and silence. Nobody knew. Only the Dulligan people would converse with them. In now times, they're not actually extinct, but they hide quite a bit. When they're on Earth they hide…but they go up back and forth to whichever planet they go to. And they are the teachers, actually, and they pass down to us human beings the knowledge and they help us.[11]

This shared information is not primitive folklore, but actually advanced knowledge passed down over generations that reveals a lot about the Sasquatch-type Yowie in Australia.

Many people think the Bigfoot/UFO connection is merely a passing phenomenon. In the Solomon Islands and Guadalcanal, investigator Marius Boirayon tells of different races of giants, with UFO activity side by side—being observed and going on for *several centuries* according to the natives. His book *Solomon Islands Mysteries* (2009) tells the ethno-history of the island people and what they have endured over the years. Stories gathered from the elders about their villages being terrorized by the Sasquatch-like giants were classified by Boirayon as "folklore" even though the villagers accepted them as authentic history, which it most likely is. Also UFOs were a common sight there and, from what Boirayon gleaned from the islanders, there is indeed a hairy giant/UFO connection. I recommend that interested parties visit Marius Boirayon's website at *www.solomonislandsmysteries.com.*[12] The book is a worthwhile read.

The Steen Mountains of southeast Oregon look like the last place for Sasquatch to live. Yet the giants are incredibly adaptable. This region has very little vegetation except for a few mini-oases made up of springs, small streams, and clusters of aspen. There is no thick forest for cover. However, there are extremely rugged, rounded peaks—massive rock formations that suggest numerous caves. The higher one goes, the more it becomes a vast, barren mass of geography for miles around. And there are desert hot springs beyond the last eastern precipice of the Steens that are closer to Idaho. This is a unique landscape of Oregon that most people never see. My associate Jeremy Lynes, from Atlanta, Georgia, and I explored that region in July1987. As we were approaching the base of the mountain range in a van, I was impressed by the fact that a Starperson was telepathing to me, commenting on the conversation that Jeremy and I were having at the time. It was late in the afternoon and, by the time we had driven halfway up to the nearly 10,000 foot summit, it was dark and we were ready to camp for the night. We were fortunate to locate a small, level shelf with lots of trees and greenery. Just as we stopped the vehicle, a Sasquatch began "talking" to me. He said he lived in a cave on the other side of the ridge and was busy helping the Starpeople. He indicated that we had chosen a safe place to stay for the night. "I'll see you tomorrow at the top," was how he ended our conversation.

We had started taking our equipment out of the van when an entirely different voice of an ET said, "Turn around, my friend." When I did, I saw a huge, orange globe right on the horizon—a glowing spaceship! Jeremy also watched it for a few minutes as it began to slowly move, then completely blipped out. It was gone! The following morning we packed up our gear and headed up the steep gravel road toward the summit. It was 11:30 a.m. when we arrived at the top. The vista was spectacular! We had only been there less than a minute when that all too familiar Sasquatch "aroma" could be smelled. If we had had guns or ill intent, no doubt he could have released a powerful stink-bomb. I don't know if the creatures can release it deliberately or if it's merely an instinctive response to danger. Then a voice in my head said, "I told you I would be here."

Here was another example of a psychic Sasquatch with a UFO connection. If the reader is curious as to why I seem to have so many interesting encounters with communication, the answers are:

1) they chose me, I didn't choose them;
2) I am a world traveler and enjoy experiencing new horizons with an open mind;
3) I camp out frequently during summer and winter in isolated areas where there is UFO and Bigfoot activity; and
4) my professional speaking and writing skills are valuable to them in educating the public at a time in the Earth's history of man when our planet is literally dying!

Each one of us should look beyond the McDonald's arches, sports bars, and Nintendo games to figure out what we are really here for. I believe that we all have a purpose—we all are valuable human beings with gifted skills and specific jobs to do. But, it's up to us to find them! As Edgar Cayce said, "We are all corpuscles in the blood of God," and we are in physical bodies to act out and express a part of God on this plane of existence.

When I was on a three-week expedition to southeast Oklahoma and western Arkansas in March of 2006, I was personally directed by an Ancient One to a specific place to have a meaningful experience. I was alone as usual. It was cold and snowy, so I hunkered down in my tent reading and writing most of the time. In spite of inclement weather and roughing it, everything was hunky-dory. The encounters and data I collected made it worthwhile.

Nothing happened the first few nights except cold damp rain. Then just as I was falling asleep on the fifth night, I was alerted by the sound of a long, low growl! My heart started to pound since it was no more than seven or eight feet from my tent. My first thought was of a bear. It growled again. I became nervous, because once a cougar had come to within twenty feet of my tent early one morning, and another time, a bear kept pushing its nose against my tent while making sniffing sounds. When *it* growled a third time, whatever-it-was had not moved. I was baffled.

"Dream man," a female voice telepathed and went into a slight giggle. It was Haloti. She and her male partner had traveled over one hundred miles in two days to be in that region. I tried to find out why, but she did not answer me.

Haloti's clan affectionately refers to me as "Dream man" because they say I am a visionary who is ahead of my time. They were deliberately playing with my head just to have fun. Usually we interact in the Kiamichi Wilderness, but she was checking up on me to see if I had found the area that she suggested. They didn't stay, but moved off within a minute. I still don't know how they spend their time. The beings are always busy going somewhere, and they seem to be working with Earth energy areas that we can't feel or see as humans. So I still have a lot to learn and discover.

My last night there, I was waiting for Pushoma to visit me. He was the chief of all the clans of the Ancient Ones. Through Haloti, I had asked him to take me into another dimension. Her answer was vague, yet directed me to a very unfamiliar place in the forest. She knew some of the names that White people gave these areas on a map. Then she would add her own observation of nature by saying that the area I was seeking is on a hill, overlooking a river, toward the setting sun—in an almost Native Amerindian way, a nature person stating all the indigenous signs as a reference point when communicating.

That last night it was 28 degrees, and I had wrapped myself in my sleeping bag. I sat looking at the clear night sky wondering what might happen next. Sometimes, when frigid weather hits or the rain won't let up, I start questioning just what I am doing "out here" instead of being in a warm, comfortable bed. But the freedom to explore one of life's mysteries and to get "pulled" into another world makes hard times more rewarding. It was always that risk I took! It would all be worth it to meet Pushoma when he came, and to hopefully experience another quantum realm.

Then I noticed an airline jet fly over. I was daydreaming and came to realize that there were two round "lit" flying objects that were barely moving between the hills, very low. As I perked up and looked around, there was another glowing sphere right at tree-top level, approximately 60 yards behind me. Then another commercial jet was cruising at least a mile out. It didn't resemble the behavior or shape of the UFOs. After three and a half hours I was too cold to look any longer. During that time I had counted 12 spaceships! I used to keep track, but I have seen about 175 ships or more in my life. The reason? I have spent most of my life in the forest and wilderness areas all over the world. I *always* go out to look at the night sky. Even when I drive during the day, I see UFOs and feel blessed to have experienced so much in my lifetime. I look up at the sky all the time. Even though I had the privilege of viewing spaceships, I was disappointed that I had not met Pushoma.

The next morning I was to meet my associate Peter, who was taxiing me between wilderness areas once a week. There were more places to investigate. I was anxious to tell him about the spaceships that I had seen. But when he arrived, Peter was elated about an encounter that he just heard about at a convenience store at a village several miles west of my camp. When he stopped for cigarettes at Talihina, he heard a group of town folks chatting up a storm. Several people had seen a UFO go over the village at 4:30 a.m. Even the clerk in the store had stepped out to watch it.

To make things more interesting, two men arrived (whom apparently most folks knew), who had been out laying out corn feeders one week before turkey hunting season began. They said their camper was parked up in the forest where they had slept overnight. The two woodsmen woke up at 4:30 a.m. and built a fire to make coffee. At 4:45 a.m. the guys were dumbfounded to see a large round spaceship at tree-top level, barely moving, but coming straight towards them. They continued to observe the object until it stopped directly above them, turning a huge beam of light down on the two men. At this point (as the two had men said), they were panic stricken! All their equipment was quickly thrown in the back of the camper, the coffee was dumped on the fire, and they got in the cab as quickly as they could. The vehicle went speeding down a winding, muddy road. As the men rounded a curve, there in the middle of the road was a ten-foot-tall Sasquatch, with a shorter one standing at the edge of the woods. Before the driver could stop to avoid hitting the creature, it turned and looked directly at them as they braced for impact. But the Sasquatch immediately faded, so the vehicle drove *through* empty air! It had dematerialized, and they started going crazy at that point. Apparently the being had had enough time to "blip out." If he had been walking across the road and not seen the truck in time, he might have gotten clipped or killed, having been "caught" in this dimension.

I asked Peter if he had asked the men for their names and phone numbers, but in the excitement he had forgotten. Peter did remember the name of the road and place where all this occurred. Strangely, it had happened less than one hundred yards from my campsite. After breaking camp, we drove over to where the would-be hunters claimed to have had the encounter. We parked Peter's truck and walked on each side of the road looking for tracks. There, on the soft mud, were several 14 inch and 16 inch tracks with five and six foot strides. Plus, skid marks could be seen where the camper tried to stop. No one else was up there at that time of the year, so it was easy to read the signs exactly as the two men described. They were telling the truth! It had happened five hours before we arrived on the scene. We took photographs of the clear, perfect tracks.

From there, that same day, Peter drove me to the Black Fork Mountain Wilderness in Arkansas. I studied the map, looking for the area that Haloti said has a busy vortex. Soon we parked the truck at a trailhead and began hiking into the forest. As soon as we entered that stretch of timber, Peter's eyes became wide. He looked at me and said, "They're here!" A powerful *wave of energy* came over us, along with a feeling of love and a sense of security. Our entire bodies were vibrating at a high rate with goose bumps all over. This is normal when the man-creatures are present.

At a remote spot I located the vortex, then set up camp about one hundred feet or more away. The reason for the distance was to show

Author examining tracks in the spot where the truck drove through the dematerializing Sasquatch in southeast Oklahoma near the Arkansas border

respect and not to disturb the Sasquatch's normal activity. This was their sacred space. We honored that space when I laid tobacco and asked permission to camp there. The beings said they knew me and felt comfortable with me in their area. Peter left me to camp alone and would return in one week.

The first five nights were "normal." An occasional strong wind blew through the pines with creaking branches that disturbed my sleep from time to time. The sixth night was different. At 9:50 p.m. the portal became busy. I could hear a low gravelly "voice" talking an unintelligible language. It was a sound I had frequently heard in other places where the giants dwelled. There were two separate tones, making it obvious to the listener that there were two separate creatures. At a distance, in another direction there was a powerful bellowing hoot. Later the same hoot was heard coming from yet another direction. Then there were more low mumblings from the vortex. This went on for about an hour and a half. I telepathed asking them to take me with them through the invisible "door," but they were sticking to their secret agenda. There was no reply. The next day Peter picked me up. The area had not been as fulfilling as I had hoped. We drove 160 miles to the place where I spent the third week in a rough cabin in the Kiamichi Wilderness in Oklahoma. During that time, I heard twig-snapping and someone walking around outside at night, but nothing significant happened.

Over the years, I have dowsed different locales where Sasquatch were working with friendly ETs. Sometimes the sites are two or three thousand miles away, but the investigator experiences them nevertheless. In January 2006 I did this for Mary Rau, whom I had not met when I did it, but I felt she could be trusted. Mary was a skeptic, but not for long. She bravely followed my instructions and went camping in southern California's high desert *alone*. Mary chronicled the event as occurring on Saturday January 21st:

> Many commercial jets were seen flying over, and at 6:15 p.m. I spotted a UFO flying right over my head. My jaw hit the ground and I thought I was hallucinating. I swear I had had nothing stronger to drink than green tea. It was circular with a strange pattern of pinkish-red and lime-green neon-like lights underneath. I continued to watch in disbelief as it glided soundlessly to a nearby ridge where it hovered for nearly an

hour. At 8:30 I was huddled close to my small campfire when I heard bipedal footsteps crunching towards me.

Mary quickly turned on a flashlight instead of waiting for the Sasquatch to connect with her. Her fear factor made the creature withdraw, as not to heighten her apprehension. He was being courteous and trying to sensitize her to himself and the ships.

Then, in July of the same year, I had the pleasure of meeting Mary when I invited her to Washington State to meet my neighbors—more friendly hairy folks! She camped in the forest in back of my place and made herself at home. On the second day of Mary's visit she documented:

> On July 24, 2006, at 1:10 p.m., I was sitting on the deck of Kewaunee's cabin reading a book when I heard a loud *snap* of a tree branch. I looked up and about 60–75 feet from me in the woods stood a Sasquatch.

Around my place, one must be careful what they wish for. The creature then became invisible, so Mary walked down my driveway to the dirt road and stood looking into a thicket of trees and bushes. She continued:

> I telepathed, "My name is Mary. Are you a Sasquatch? What is your name?" Before I could complete my thoughts, he responded and I asked him to repeat. He said, "Elin" [pronounced el-een]. I asked if I would see him at the lake the next day where Kewaunee and I planned to hike to, and he said "yes." I thanked him and went back to the deck of the cabin. A minute later I heard a branch snap again and felt that Elin had more to say to me, so I walked back to the same spot. I said "Do you have more to say to me?" He responded, "You are welcome." I burst into laughter that he had made me suffer the hot walk just so he could respond to my thank you.
>
> Back to the deck and glorious shade, again another branch snapped. At this point, I decided to just respond from where I was and asked, "Do you have more to say to me?" I suddenly felt a rushing sound in my left ear, as if it was filling with water. I touched my ear and it stopped. I asked, "Did you do that?" Elin responded, "Yes." I then asked, "Are you playing a game with me?" He replied, "Yes." I laughed and said, "Enough already. I'll see you at the lake tomorrow."

Mary did see Elin across the lake observing her, and heard loud "beeping" noises at 4:45 a.m., and later a "tremendous *whoosh* from the lake," and an oily film after that, that had not been there before, plus a noticeable drop in the water level of the lake by two inches! It gave her a lot to think about.

In August of 2010, I received a letter from a man named Andrew Robson from Oregon. He was interested in the Sasquatch people because he could hear them walking around him when hiking in the forest. Andrew had read my book and wanted to correspond with me. I soon realized that he was an excellent subject for telepathic communication, so I taught him. A couple of weeks later he encountered a Sasquatch that ran on all fours. Another couple of weeks after that, he saw a large light-grey Sasquatch walk into the woods at fairly close range. Then a week later a female Sasquatch briefly spoke to him. Andrew was thrilled! Eventually a dialogue began between him and a male named Oea [pronounced Oh-a-ah in the Sasquatch language]. Though contact began in the forest, soon Oea would appear in his apartment, interdimensionally, often with three small ETs! Oea told Andrew the following message on September 21, 2010, and gave permission for it to be included in my book:

Contactee Andrew Robson

Halito,
I am the Grey One—Oea.
My home is the hills and mountains.
My Tribe has been here for many hundred years.
My Mother was of the Ancient Ones
My Father is of the Bukw's
My ancestors were the C'iatqo [pronounced Sea-at-co]
And the Amorites—
Many Tribes have come from Many Lands.
The Star People brought my ancestors here long ago from many lands.
Many Races are connected.
I have come to give you Deep Meaning—
I have heard your thoughts—
I have come to tell you of the signs you have witnessed.
A Great Spiritual Shift is coming to the Earth Mother—
This is the "The Harvest of Earth"—The Sacred Purification—
You are being led out of the Darkness;
Into the Light.
The Star People have bases under the seas to balance the
Earth Mother as the Celestial Conjunction Dawns.
Many white soul entities are coming into this Dimension
to Guide the way.
In the forest and within the Earth Mother will God's Chosen
find Solace.
Many mysteries will be revealed.
Many souls will be awakened.
You must Act with the HIGHEST LOVE.
The Purest heart must beat within You.
My People offer to you the Love Stone Amorite…
These stones have the power to connect to the laws of Love.
These are very sacred Stones that my Ancestors have known
of for many years.
Kewaunee knows of this stone.
We have given you both this message.
It is of GREAT IMPORTANCE that you begin to experience
and manifest the New Energies that are now coming onto
the Earth Mother.
Fear NOT, For we are with you!
We send you with Hope,
And Blessings
from D'sonoqua

Andrew was thoroughly elated that this profound message was given to him and the author pertaining to Earth changes and 2012. These ongoing visitations have dramatically changed Andrew's life. He says it has given him new meaning, and he views the world around him in an entirely different way. The materialism and social status in our society have become insignificant to him since having these spiritual encounters. One time when Oea appeared, Andrew was reading a book on his bed. Oea stood at the foot of the bed, asking Andrew to take notes because it was important information that he was going to share. The notebook was on the bureau next to Oea. The being turned and looked at the notebook. Andrew was astonished to see the notebook levitate and float across the bed to land gently in front of him. (He is the third person I have documented who claims levitation is associated with the Sasquatch people.)

Another time Oea told Andrew that he could not talk for very long because he was on a mission and two more of his people were on a spaceship above them. Once, Andrew telepathed to the Sasquatch and three spaceships showed up, because the hairy-folks were inside the craft. Still other times, unusual-looking ETs would appear and talk with him in a loving way. One being said Oea had sent him, then asked to see

The kindly interdimensional Oea

Andrew's Amorite stone. He handed it to him. The strange but kindly 6-foot-tall being "held it in his palm and closed his eyes momentarily. Then when he gave it back, it was very warm and glowed in a soft silver light."

The last time he was visited, it was by a 3-foot-tall grey ET with a silver suit. The being suddenly materialized while Andrew was taking a bath. It appeared to scan his body, gave him an important message, then dematerialized. Andrew was so impressed by the ETs kind message that he quickly said mentally, "I love you."

There are many different races of greys, just like there are different kinds of Asian or Amerindian tribes, who have their own specific beliefs and culture. These greys apparently don't abduct people; they have a different agenda. Like my own ET visitations, Andrew's are innocuous, insightful, and interdimensional in nature. I am still mentoring Andrew, as requested by Oea, so that he follows a certain protocol that's comfortable for these beings when interacting with Earth people. And if all this sounds like science fiction, it's not! Welcome to the *real* world of Sasquatchery.

A human being is part of a whole,
called by us the Universe,
a part limited in time and space.
He experiences himself,
his thoughts and feelings,
as something separated from the rest,
a kind of optical delusion of his consciousness.
This delusion is a kind of prison for us,
restricting us to our personal desires and affection
for a few persons nearest us.
Our task must be to free ourselves from this prison
by widening our circle of compassion
to embrace all living creatures
and the whole of nature in its beauty.

Dr. Albert Einstein

CHAPTER 4

THE HYPERDIMENSIONAL SASQUATCH & QUANTUM PHYSICS

A man approached me after I gave a lecture at a conference in San Jose, California, in August of 2007. He said that while elk hunting in Colorado with his 18-year-old son, a most startling Bigfoot encounter had occurred. While deep in the forest they decided to separate, but he warned his son not to wander very far. He told him to find a good spot and stand quietly by a tree. Soon they were out of sight of each other and in position if a deer or elk moved in their direction.

After a time, the teenager heard something bounding toward him. He was leaning against a tree, and suddenly a frightened deer came dashing past from behind him. Before the young man could collect his thoughts, a pounding of loud footfalls ran by him—close enough that he felt the breeze from the creature that made them. The problem was that there was nobody there—at least nothing visible!

The confused deer was grabbed by an invisible force and violently thrown against a tree, instantly breaking its back! This freaked the kid out. He literally did not know how to react as he stood there in a state of awe and fear! His father told me that the Sasquatch never materialized, but picked up the dead deer, dangling it by its hind legs towards the boy in the same way that someone holds up a dead rabbit after its been shot. The creature held it easily, as if it only weighed a few pounds.

The giant stood in front of the teenager, but was physically indiscernible except for the deer dangling in mid-air. It appeared to offer the animal to the boy as the Sasquatch slowly moved toward him in a non-threatening way. For a moment he thought of shooting at where the Sasquatch's body should be, but changed his mind. He was scared and in a state of confusion! It was too much for him to comprehend, so he turned and ran while yelling for his father. The father told me he figured it was a Sasquatch, as he had read about the possibility that they were interdimensional. He said it took a while to calm his son.

Knowing what I have experienced and having had lengthy conversations with others who had invisible encounters, I had no problem with this man's veracity. He was very anxious to share the encounter since the first researcher with whom he spoke had laughed and rebuffed him. Conferences are great places to meet other witnesses and people with similar interests and experiences. But what are we to make of this type of encounter? It's not normal for the young man, but is very normal for a Sasquatch. I can't recall if the man said there was an odor. My years of exposure to the psychic Sasquatch tells that me everything points to interdimensionalism. There is no other creature in the North American forest that could do what his son described. Yet how does a being that looks so primitive and animal-like possess the ability to alter sub-atomic particles and exhibit such mastery over its physicality?

Perhaps part of the answer lies in what they have told me and other contactees—that originally they evolved on another planet and were brought here eons ago by ETs! On our planet, societies are so immersed in their own microcosm of a world that they have little time to contemplate the Big Picture either physically around them or in other dimensional planes. I am totally convinced that Darwinism, spiritual evolution, interdimensional planes of reality, and ET interventionism are all happening simultaneously to produce a living universe. As one ET told me in 1984, reincarnation is real, but more expanded than the way some people believe it. He said it is really all one life as we recycle in and out of different bodies, on different planes, in different lifetimes! So in a sense, we never die. Or as Einstein said, energy cannot be created or destroyed. Invisible dimensions are an integral part of this energetic cosmic process.

In Fred Beck's classic book, *I Fought the Apemen of Mt. St. Helens* (1967), he tells his personal story of the now famous attack on their pine log cabin by a group of Sasquatches in July 1924 at "Ape Canyon." In the short monograph, Beck relates in great detail exactly what happened to him and four other gold miners when they were working their claim in the southern region of the Washington Cascades. It's a fascinating account that is well documented by a tell-it-like-it-is old-timer. The important point to emphasize is that Fred Beck is the first person on record to specifically state that the "apemen" were indeed *psychic*! Fred Beck said, in his book, that he and his prospector companions were *all in agreement* "that we were dealing with supernatural beings."[1]

In all the renditions of this event in books and magazine articles I have read, I have never seen any author mention this vitally important fact. Mr. Beck emphasized that he and four other witnesses experienced the psychic Sasquatch. Just because the psi realm as a subject was not in vogue doesn't mean it didn't happen or wasn't real! The primary key to this entire phenomenon *is* the anomalous reports! At this time some disciplines of science are seriously investigating the psychic realm, and as a responsible investigator, I must include the interdimensional Sasquatch with telepathic abilities.

Anything in nature is up for grabs to be investigated and in our universe anything is statistically possible. In a few years people will look back at the purely "flesh-and-blood" researchers and laugh that it took so long for them to discover the psychic Sasquatch. These clever interlopers *are* flesh-and-blood, but they somehow have the ability to go from our dimension into another! What Fred Beck articulates in his book is an historic account. It already happened and cannot be changed! Documentation like his is valid and insightful, and only continues to reinforce what I have already learned from experience and what numerous witnesses continue to voice. The psychic aspect to the phenomenon is being irresponsibly overlooked.

One of Beck's adamant statements was:

> No one will ever capture one, and no one will ever kill one…

> in other words, present to the world a living one in a cage, or find a dead body of one to be examined by science.[2]

Also, he added:

> First of all I will say that they are not entirely of the world.[3]

Very sagacious for someone with no formal education. Why was he so tenacious when making the remarks? Because of what he and his four comrades encountered over a six-year period in the wilds of the Mt. St. Helen's area nearly ninety years ago:

> These beings bear a direct association with the psychic realm," he said.[4]

Fred Beck continues to describe evidence that the Ape Canyon creatures were supernatural from an interdimensional standpoint:

> Another very striking experience which shows that they cannot be natural beings with natural bodies: It was before we made our cabin, we were staying in a tent then. The tent was below a little cone-shaped mountain called Pumy Butte. A little creek flowed nearby, and there was a moist-sandbar about an acre in area. We would go there and wash our cooking utensils and bring our drinking water back. Early one morning Hank came back to the tent. He was rather excited. He led us back to the moist-sandbar, and took us almost to the center. There in the center of the sandbar were two huge tracks about four inches deep. There was not another track on that sandbar!
>
> There we were standing in the middle of the sandbar, and not one of us could conceive any earthly thing taking steps 160 feet long. "No human being could have made these tracks," Hank said, "and there's only one way they could be made, something dropped from the sky and went back up."[5]

In 1984, along the Sprague River of eastern Oregon, I followed Sasquatch tracks in three inches of snow that abruptly ended! It confused me back then. The Indian woman who lived with her husband at the edge of this wild region said she had found the "disappearing tracks phenomenon" twice before. Then she told me that when the Sasquatch were visiting the area regularly, she wrote to Canadian researcher Rene Dahinden. He arrived with a camper and stayed for two weeks. While snowmobiling, she discovered two sets of tracks in the snow, large barefoot Sasquatch tracks and tiny boot prints walking beside them.

Both sets of tracks ended in the middle of a clearing, showing that the beings who made them had simply disappeared. When she showed Dahinden the tracks and where they vanished, he merely walked away and changed the subject, she said. The woman told me she was baffled by the disinterest in analyzing what she considered rare and significant evidence of interdimensionalism that tells the full story in the snow. Is this a case of a "cover-up"? Are mainstream researchers consciously or unconsciously concealing the evidence?

Fred Beck was so evolved in his thinking because he was a kind-hearted spiritual person. His little booklet is a gem of information—the interdimensional experiences were truly objectively recounted. Beck says of Bigfoot researchers in recent years:

> The expedition now is a good thing…let the young ones explore, nothing could be better than for them to try and solve one of life's little mysteries. It should in time lead them to the gates of psychicism.[6]

It seems very clear to me that the average researcher has an intrinsic problem with aberrant states of reality, otherwise the Sasquatch mystery would have been solved a half century ago. That's because there has never been the application of an interdisciplinary approach. Zoologists and animal experts were looking for an animal that doesn't exist!

Quantum mechanics has presented various theories and complex laboratory experiments in an effort to answer the "invisibility" problem. England's Ministry of Defense revealed in October 2007 that they have devised new technology that can make tanks invisible. The British Army said they can make a vehicle completely disappear—James Bond style! It is "predicted that an invisible tank would be ready for service by 2012."[7]

I am not comparing a tank to a Sasquatch, but merely illustrating that *technological* progress is being made, whereas the giants can either control atoms in cells of the body or effectively cloak themselves, apparently using telepathic suggestion. As a trained hypnotherapist, I know that with a hypnotic suggestion one can tell a person that there is no one else in the room and the subject will believe it, since they don't see anyone. This may be analogous to the psychic dynamics of a Sasquatch mind with invisibility.

Somehow, I feel it's far more complicated than this. The psychic Sasquatch seem to think themselves invisible, as I have observed twice.

Also I have twice witnessed a spaceship dematerialize, and one of those times I had an awe-struck witness with me to prove it. He stopped poking fun at me after that. I am not a physicist and make no claims to understand quantum mechanics, but science is certainly on the threshold of proving and demonstrating invisibility.

This vital concept in relationship to Starpeople and Sasquatch is being reported so frequently that a good investigator needs to look at all possibilities, especially toward where the evidence is leading. The keys to the entire mystery, based on my multiple contacts in the last three decades, will be evident if a person:

1) has an open heart that projects unconditional love by letting go of all fear and negative emotions;
2) becomes erudite in mental telepathy;
3) accepts the fact that the Sasquatch people, ETs, and *some* out-of-place cryptids *are* interdimensional; and
4) accepts that many or all of these physical beings have access to "portals" or "vortices" that lead to another dimension or parallel world.

Personally, I don't need a quantum physicist to prove anything to me, because I have been busy experiencing these phenomena—over and over again! I do want to be responsible by sharing it with the public at large as well as pointing other open-minded and responsible investigators in the right direction, based on my findings.

The New Journal of Physics in December 2008 had an editorial on "Cloaking and Transformation Optics." Here they quote the late Arthur C. Clarke: "Any sufficiently advanced technology is indistinguishable from magic." [8] It's extremely complex, but in short, science now has the potential to transform optics to produce invisibility.

The National Geographic News puts it more simply by explaining that such an invisibility cloak has "a way of bending the geometry of space so that light from all directions travels around an object, rather than hitting it." (December 10, 2008) Then there is the new photo-stealth camouflage technology from Military Wraps, Inc., that is used to conceal vehicles and airplanes in a specific terrain. This concept interests me because many Native Amerindian friends and acquaintances have told me that a Sasquatch can make you think it's a tree, when in fact the creature is standing right there. It took me a while to accept that.

In August 1999 in the Mount Adams Wilderness, I had telepathic communication with a friendly male and female Sasquatch. Then they stopped conversing. Some thirty minutes later I began hiking out. Just a few feet from the trail head I walked into a zone of strong "psychic" energy. When I stopped walking, I could feel a "wave" of Sasquatch energy. Upon looking into a grove of conifers, the energy intensified, but all I could see were trees. The couple began to talk to me and were happy to meet me. Their attitude was very playful and exuberant. For the first time, I was convinced they somehow altered my vision, creating an optical illusion. I can't prove that; all I can do is convey what occurred and what I felt.

I am convinced that what I am sharing in this monograph has significant value in better understanding the quantum Sasquatch. As Einstein said, "Reality is merely an illusion, albeit a very persistent one." He also reminds us that, "The only real [sic] valuable thing is intuition." And most intuition is based on one's feelings, which scientists insist on repressing so they don't interfere with their "objectivity." However, conducting Sasquatch research is not like studying a North American mountain gorilla, because *they* don't exist. I keep emphasizing the importance of the Sasquatch's psychic ability, since we are dealing with forest-dwelling ETs of sorts, who tell contactees they were brought here before humans were brought here. They do not waver on this fact. We are the underdogs when they are present, which is why no one can study them in a classical ethological sense. It is really they who are studying us. So it behooves us to let go of our arrogance and make friends, since we will never have control over them no matter how sophisticated our technology becomes.

In 1979 I began experiencing nearly the entire array of psi phenomena. At times it overwhelmed me. I didn't understand that my kundalini was opening. During that time, I met a man named Tom who possessed a high degree of psychic ability. He practiced assiduously using several Eastern techniques to develop and enhance these powers. The man was humble and matter-of-fact about the different abilities he demonstrated to me and others. Being a neophyte and coming from an extensive background in the social and biological sciences, I was easily impressed—never having met anyone like him before. One day Tom sincerely told me that he acquired the ability to become invisible. I didn't believe him! Apparently he had developed and expanded on a technique to use visualization to speed up his body's vibrational frequency.

He likened the idea to the rotating blade of a fan, If you turn the fan on low, you can still visibly see the blade slowly moving. But if you switch it on high, the blade virtually disappears from sight, yet the physical blade is really still there. Tom claimed it worked.

He told me that one day when his fiancé was visiting, he had asked her to sit across the room from him in silence and then tell him what she saw. After several minutes of concentration he dematerialized and his fiancé screamed. She was totally freaked out! Then Tom slowly came back into view again after "turning the fan on low, then off." While he was telling me all this, his fiancé came over and listened in. When he finished sharing his story, she turned to me and confirmed that what Tom told me was indeed true. So there are gifted people in this world, and then there are the super-ultra-gifted psychic Sasquatch.

Native American Arla Williams of Oklahoma told me she is being mentored, since August 2010, by Keshema, a Sasquatch who is a member of the Council of Elders. In December 2010, while she was walking in the woods with him, a flock of wild turkeys scattered in front of them. Arla decided to ask Keshema if he could make her invisible so she could approach the birds without being seen. To this he agreed. Instantly she was in another realm where she could see the turkeys, but they couldn't see her. Arla was impressed when she walked up to them without being noticed. She confided that she felt strange after returning to her normal state and that the between-worlds "spaciness" stayed with her for a while.

Physicist Dr. Fred Alan Wolf discusses an array of related subjects that is pertinent to Sasquatch behavior in his books *Parallel Universes* and *Taking the Quantum Leap*. There is plenty of substance in the quantum arena, including the research of the brilliant professor of theoretical physics, Dr. Michio Kaku, a cofounder of the string field theory. His books are *Parallel Worlds*, *Hyperspace*, *Beyond Einstein*, *Physics of the Impossible,* and so on. I am mentioning these for more advanced and prolific readers who want more data about invisibility, teleportation, time travel, etc. Some of this material strains the imagination as well as a few neurons. But the books impress upon the reader that the 21st century is advancing into the realm of near–science fiction technology.

Our government may already have ET-type technology, possibly secretly acquired from some race of Starpeople, that would shock the average person if we really knew. Many UFOlogists claim they have spoken to government whistle-blowers who have made fantastic claims concerning this. True or not, we are publicly making advances in leaps and bounds, if only we would use it for the higher good of all concerned. The military is always interested in anything that will help it be one up on a potential enemy. Jim Marrs' book *PSI Spies: The True Story of America's Psychic Warfare Program* is a good one for skeptics to better brush up on their awareness to the reality of psychic phenomena.

Then, for the ultra-skeptics who still don't believe in a psychic Sasquatch, there is a truly mind-expanding book by scientist Dean Radin, titled *Entangled Minds: Extrasensory Experiences in a Quantum Reality*. Dr. Dean Radin is the Laboratory Director at the Institute of Noetic Sciences in Petaluma, California. For two decades he has conducted research on psychic phenomena at Princeton University, University of Edinburgh, and the University of Nevada. Plus, at SRI, he worked on a highly classified program investigating psi phenomena for the US government. Dr. Radin's book, *The Conscious Universe: The Scientific Truth of Psychic Phenomena* won the 1997 Book Award from the Scientific and Medical Network. The book has been translated into eight languages.[9]

Lastly, I will sneak in one more scholarly book of knowledge called *Interdimensional Universe: The New Science of UFOs, Paranormal Phenomena and Otherdimensional Beings* (2008) by Philip Imbrogno.[10] One might expect these to fit into a bibliography, but I mention them here to impress upon the reader the seriousness and extensiveness of psi in the laboratory as well as university setting.

Dematerialization means "something becomes invisible while it is still in the same place." *Teleportation* means "something disappears because it is projected from one geographic place to another."

In February 2006, I was preparing for a field expedition to Oklahoma and Arkansas. Before leaving I was shopping—collecting supplies and equipment—and my friend Tara was accompanying me. First, I want to mention that I have on numerous occasions asked both the Sasquatch and ETs to afford me interdimensional experiences, so I can better understand the phenomenon firsthand. The following account is, perhaps, an answer to that request. While driving due west on a familiar road to an army surplus store in Lynnwood, Washington, without prior warning Tara and I were teleported approximately five or six miles north so I ended up driving *east* on a totally different road in a different part of town. I was completely discombobulated and so was Tara—my witness to this blink-of-an-eye transfer! One second I was telling Tara that we were almost to the store, and I recall saying it was about 75 yards around the bend, and the next second we were somewhere else, miles away. It took me a few minutes to get my bearings. Suddenly I was going down a steep hill driving east, when a second before I had been a few miles away on a level road driving west! There was no telepathic communication, but it had to be the Starpeople that did it. The crazy part is, I was in heavy traffic with cars all around me. It made me wonder if anyone saw my red Honda "blip out"! Was any other vehicle transported interdimensionally along with mine? There was no way of knowing. I do know this, I had to enter the freeway and drive back to the army surplus store several miles away, all over again.

Recently, it was announced in June 2009 via *The Daily Telegraph* that a team of researchers at Australian National University had developed an entirely new method of transmitting certain data that is one step away from establishing teleportation technology and super-fast quantum computers. This is done by using a process called quantum entanglement. It is "a process in which two objects are linked together in such a way that any changes to the properties of one can be measured from the other regardless of the distance between them." So why would they spend years developing something that's not real? Because teleportation *is* real; ETs and Sasquatch use this mode of transportation all the time. Other percipients I have interviewed have seen the results of teleportation, a few of them have seen it multiple times.

In my first book I related an incident in which I telepathed to the ships flying over my cottage in Oregon asking for a healing when I had a ruptured disk and was unable to walk. Two nights in a row, while I was lying in bed, three little ETs (not greys) teleported into my room. A year later, an entirely different ET—who looked similar to a human but with a larger cranium—materialized before my eyes from a speck of light. He had also teleported from a UFO, as well as I can describe. It's literally like Scotty in Star Trek beaming someone down.

Dr. Christopher Monroe of the Joint Quantum Institute at the University of Maryland stated in the January 23, 2009, issue of the journal *Science* that they have succeeded in an atom-to-atom teleportation of information at a one-meter distance. Thus far, all the races of Starpeople who have visited me in my home were either teleported or possibly could do so using their minds.

When I was 15 years old, I joined the high school track team. I ran the mile. At one of our out-of-town meets, each team was sizing up the other, strutting around a bit as young kids do. As we were preparing for one of the races, a puny, goofy-looking kid with freckles and buck teeth stepped up to the line. Our team laughed and made derogatory comments amongst themselves. I was thinking to myself, "What's he doing in this race?" since he *looked* like a poor contender. But once the race started, the doltish looking kid became the star runner. He outran our best runners by 20 to 25 yards or more and won by breaking all the records! It was the very last time I ever judged a person by his/her looks. My point is: don't underestimate the fantastic powers of the psychic Sasquatch just because they resemble an animal.

There is an excellent NOVA DVD series titled "The Elegant Universe" (2003) about superstrings, hidden dimensions, and the quest for the ultimate theory. It's an intriguing intellectual adventure that's superbly orchestrated for television. The show is hosted by Professor Brian Greene of Columbia University Department of Physics. He is also the bestselling author of the book by the same title. Though admittedly, Greene speaks only about "what if" theories, the computer-animated illustrations give a feel for what it's like to experience another dimensional place. None of this is proof, but it's stimulating enough to have some application for a psychic Sasquatch, even though I already know unequivocally that they are interdimensional.

In October 1998 I was a guest speaker at a conference called "The UFO Experience," which was held just down the street from the Yale University campus. The keynote speaker was Astronaut Dr. Edgar Mitchell, who was a part of the Apollo 14 mission and was the sixth man to walk on the moon. He is the founder of the Institute of Noetic Sciences, which is dedicated to finding scientific answers to psychic phenomena.

So science has become more scientific in areas of psi, in an effort to understand our planet's unsolved mysteries. There is so much data in the quantum sciences that all I can do is reference a few of them, so that the reader might appreciate that there are several parallels between what science is theoretically attempting and the encounters that I continue to have with the psychic Sasquatch.

There have to be answers somewhere, and I am hoping eventually they will come from one of the races of Starpeople or the Sasquatch, if they feel like sharing the information. Certainly if "they" were to share some solid proof and quantum answers to their other-dimensional abilities, then the scientific community would start taking people's experiences more seriously. In the past when I felt confused about my encounters, I have cogitated: Are the Sasquatch or Starpeople devils, angels, parallel world visitors, or who they appear to be—evolved physical beings with profound paranormal (normal to them) abilities from inner Earth, outer space, or inner space. We must remain open if we are to resolve this problem. They only spoon-feed bits to me and other contactees. No one has had all the answers.

In October 1983, I visited a person whom I had never met before. I was introduced to her by my then girlfriend. While socializing in her kitchen, a Sasquatch astrally projected himself there and stood staring at her from just a few feet away. I could see his ghost-like body, but said nothing to the hostess. Moments later she stopped chatting and began telling me of the creature's presence. Later, the woman said she had never read a book about Bigfoot and had only glanced at a few newspaper articles in the past. Since I had just met her a few minutes before, I hadn't mentioned that I was a researcher or the subject in any way. I contacted her later and asked her to document her psychic encounter and this is what she sent me from her observations and clairsentience:

> The day you were here and the Sasquatch manifested in the kitchen was astounding. The person I saw was much more ape-like, which was only because of the superficial qualities of a lot of hair. I felt that beneath the hair there was a human, an extraordinary human creature. The first thing I felt when the being astrally projected itself to me was that it was very colorful. I felt it was extremely intelligent, sensitive, psychic, and extremely at one with the harmony of the Earth. I also felt, ironically, a very "schizophrenic" kind of duality. I got the feeling there was something rather sophisticated about the creature, and yet, at the same time, this juxtaposed, simultaneously, a rather primitive nature. Now, the only way I can explain it is that I felt this being could read minds. I felt this being could overhear verbalizations—in other words, as we talked it could tune into us. I felt also a sense of youthfulness or childlikeness about it; I don't want to say childlike innocence. This wasn't an innocent creature in that sense. Also, I thought that this particular creature

on some levels totally had man figured out; I mean this man-animal knows man better than man knows himself.

She also told me that the Sasquatch, using its mind-power, tried to stimulate her genitals. The implication was that the male creature was sexually interested in mating with her! Moments later, we walked into the living room where my girlfriend was reading a magazine. As the hostess and I stood talking, a second Sasquatch appeared astrally and the two beings were standing side by side observing us. We both immediately saw them. When I asked my girlfriend to look, she could not see anyone. This surprised me. Hypothetically, I feel it's like the old television rotary antennas—you can only see the picture if the antenna is aligned with the signal.

In Dallas, Texas, in 1987, while I was lecturing on the psychic Sasquatch with a UFO connection, a member of the local MUFON chapter approached me during intermission and asked, "Were you aware that there was a Sasquatch standing on one side of you and an ET on the other, while you were talking at the podium?" I told her *yes*! Usually they will tell me by saying, "We are here, my friend," and my body will vibrate goose bumps, because they have such a high vibrational frequency. At times, two or three people will independently tell me what they saw while I am speaking to a group. Of course, each time a witness reports the phenomenon to me, it only further objectifies my reality with these psychic encounters. When I spoke at the World UFO Congress in December 1991 at the Riviera Hotel in Las Vegas, two observers, separately, approached me after my talk to tell me about a Sasquatch and ET behind me. It is these beings' way of giving me moral and spiritual support. They told me that they realize how hard it can be to share data with a skeptical public and, at times, harassing Bigfoot hunters. Like I said, many different people see them—mostly while I am lecturing. One little old lady with a walker grabbed my hand when I was in Vancouver, British Columbia, and told me that she saw a pleasant-looking Sasquatch beside me with a sparse goatee, "and he wasn't at all scary," she added with delight.

Once, while sitting on the porch of a cabin in the Kiamichi Wilderness on a moonlit night, I heard someone with a "normal" human gait walk toward me. It was autumn 2005, all the lights were off, and the forest floor had leaves that were dry and crunchy. As I sat in the dark where the floor of the porch was three feet above the ground, I truly expected a regular person to come into view. The footfalls suddenly became quicker, causing the leaves and sticks to rustle and crack. The quick pace made me jump in the quietude of the night. I had a flashlight in my hand, so I spun around and shined it toward the noise that was by then five feet from me! It was an invisible juvenile Sasquatch about six feet tall. He leapt about eight to ten feet away from the porch in one jump because of my flashlight beam. Apparently he reacted as if he was visible even though I couldn't see him. He was just teasing me, trying to spook me. Immediately I telepathed to him, apologizing for shining the light in his direction, explaining that he had startled me. Also, I told him I would not turn my flashlight on again. Then he walked right back up almost in front of me in a relaxed fashion, even though I still could not see him. Again, there were lots of fallen leaves surrounding the porch, so the event was easy to hear. It was a young, friendly male that would have been eye to eye with me had he materialized. The youngster's "energy" made my body vibrate like no other situation can produce. He spoke not a word, yet clairsentiently I could *strongly* feel his joviality. The boy "hung out" with me for a while listening to my thoughts, then slowly went back into the forest.

The November 2007 issue of *New Scientist* stated that astrophysicists have concluded that parallel universes are real and valid. They are like an overlapping fold in space. No one knows what's really there unless a neighborly ET invites you through—which, no doubt, has already happened to some select individuals. I am convinced that the Sasquatch, at times, only partially enter these dimensions, but am unsure why.

For example, while visiting some friends in August 2005, on a ranch in southwest Oregon, in semi-wilderness, I saw two sets of red glowing eyes about ten to twelve feet off the ground at a distance of 100 to 125 feet. Earlier, one of them told the lady of the house that they would come to visit when I arrived. This was my first night there and we were all sitting out on the back deck waiting for their entrance. They moved around in the trees. There were no lights on, inside or outside of the house. The red glow from a pair of eyes was plainly seen. The two people who sat beside me are my witnesses. Two nights later I sat out by myself and a seven-foot Sasquatch with red glowing eyes stood by a small bush about 35 to 40 feet from me. That was the first time that I had seen the red-eyes phenomenon, and it was very spooky in spite of their friendly nature. Physicist Dr. Fred Alan Wolf refers to such overlapping of parallel worlds as "leakage" between realms.

Yogi masters in India study Vedic texts, which are a collection of ancient spiritual knowledge that came originally from Indo-European and pre-Aryan shamanistic beliefs in approximately 2,500–1,500 BC. In Hinduism, secret doctrines written around 700–300 BC in the *Upanishads* called the "Yogatattva" contain mystical philosophy and spiritual practices. Students of raja yoga were taught the use of paranormal powers called *siddhas*. One of these abilities was the practice of invisibility. In one of the earliest treatises in India, the "Yoga-sutra," there is description on how this is done through concentration and a specific type of meditation. The mentor usually possessed the definitive instructions with which he guided his student in reaching that state. This is not a trick, but a very ascetic discipline. Historically, the ability to practice human invisibility has been in existence for thousands of years. We only need to read the life and works of the spiritual adept Sai Baba, who is still performing miracles in India today. If certain yogi masters can do this, why not a more mentally evolved Sasquatch person?

While doing field research in Oklahoma, I was leaving food for the Ancient Ones and Sasquatch, as both races enjoyed visiting me at the wilderness cabin. For a couple of days, I telepathed to Pushoma, the chief of the Ancient Ones in that area. I was going to leave the following day and wanted to give him a gift of a Brazilian crystal. The crystal was placed on a wooden bench against the wall in a small, one-room cabin. My colleague Peter slept on a mattress on one side, and I was in a bed on the other, with the bench in between. During the night I heard a loud *thud*! It woke me, but I just opened my eyes without moving. I was facing the windows with a porch directly on the other side. Since there was no visible movement from one of the creatures, I assumed it had gone and went right back to sleep.

The following morning as I was dressing, I noticed that the crystal was gone. Because Peter had been examining it at bedtime, I thought he might have brought it out on the porch with his morning cup of coffee. I was wrong. He said that during the night he heard a terrific *thud* directly behind him and "felt" someone there. To his dismay, Peter was unable to move to see who was there. His body was paralyzed! Together we figured out what happened. Pushoma had come in during the night by walking through the solid wooden door. The unlocked door was squeaky enough that we would have heard the noise and woken up. But when Pushoma picked up my gift, he was enamored with it to a degree that, when he turned around to leave, he bumped his head on the eight-foot-high beam that ran the width of the cabin.

I could cite many more examples of their mutable physiology while walking through solid matter, but the final case I'll mention also took place in a cabin deep in the Kalmiopsis Wilderness in southwest Oregon in August 2007. There were three of us—myself, a movie producer from Los Angeles named Chris, and a movie actress I will call Joanie. Before leaving Washington State I had asked the Sasquatch if their friends in southern Oregon would come to the remote cabin to visit us. The answer was *yes*.

Upon arriving at the old cabin, the early evening fog was rolling in along with the coolness of the six-thousand-foot elevation. While I stood 75 feet from the building, it looked eerie—draped in a ghostly fog. (Later, a Sasquatch movie called *Letters from the Big Man* by producer Chris Munch was filmed at the cabin, for which I was a consultant.) I walked to the edge of the forest in the direction of the Sasquatch encampment and telepathed to the family of four. Immediately they sent me "visual telepathy," showing me a rather cautious, almost suspicious-looking female with a young son who was about four feet tall. The woman's appearance was similar to the 1967 Roger Patterson film but slightly more human. The large male intimated that he and his older son (just over five feet tall) would hike up to visit us. He said his son was excited and anxious to see a human up close, for they knew we were safe to be around. He told me they were three and a half miles from the cabin, and that there was only one other Sasquatch couple living in that region of the mountains, but they were a lot farther away at that time.

I began unpacking my wilderness gear and soon found a comfortable bunk to place my sleeping bag on. Then my two companions were yelling for me from outside by the picnic table. When I approached them, Joanie said she had heard a noise in the forest and a voice in her head that said, "Where's Kewaunee?" I walked over and leaned against a tree and telepathed, "I am here, my friend." I could not see the Sasquatch, but he told me that during our one-week stay he would watch over and protect us. Bears, cougars, and occasionally human intruders were the three potential dangers in this remote part of the Kalmiopsis.

My two friends were elated about this opportunity to be so close to a Sasquatch, which they were afforded because the Bigfoot people trust me to associate only with people who have a purity of heart.

I requested that the Sasquatch make a noise to wake us up at night, so my inexperienced friends would know they were there. At night, they are more active and feel secure when moving around humans. He agreed.

At 3:20 a.m. there was an horrific *slam* and ten minutes later, another. The Sasquatch let his son have fun producing all of the racket. The son had unhitched one of the heavy shutters and then slammed it hard. It reverberated throughout the entire cabin. Moments later, Joanie saw a gigantic shadowy figure move swiftly by her window. It was a moonlit night, and the reflection shone eerily though the windows on the east side of the structure. Chris reported that the playful juvenile's hand distinctly touched him on the shoulder from an invisible dimension.

During the day I looked for spoor, but found nothing. We went for hikes, told stories, read books, studied maps—basically did innocuous tasks in a leisurely way. Everything was done in a non-threatening way, just as I have done over the past three decades.

Now, the doors of the cabin were shut and locked at night. Yet every single night the youngster would walk through the solid wood. At will, he could be invisible or become physical. This was proven when Joanie woke up hearing a noise that she thought was Chris was going through her backpack. (I slept in a separate room.) When Joanie opened her eyes, her head was positioned facing downward. There, about three feet away, were a pair of very hairy legs! With the moon shining in, she had enough light to realize that it was the inquisitive "teenager." She wasn't afraid, but did not move so she wouldn't startle him. He was welcome. We enjoyed his playfulness.

The whole idea was to interact with them by simply going with the flow—without superimposing any rigid values on them. Why? Because no matter how you analyze it, they are still the ones *always* in control. Why? Because they are more intelligent and wiser than human technology. Why? Because they somehow are more evolved, while still following the rudimentary laws of nature and the universe.

The last night at the cabin, I asked the father if they would not wake us up, because we had to get up early and needed our sleep. He complied. The night before, the boy Sasquatch had made noise in the kitchen examining the pots and pans. During the night I sat up to take a drink of water, and a gigantic silhouette passed in front of the window three feet away. But an invisible finger from the mischievous boy poking my triceps caused me to jump. They love to surprise people. So this is how they are. This is what they do if a person is pleasant and forthright when engaging the hyper-dimensional Sasquatch. If they throw rocks, you can be sure they don't like your attitude!

A woman named Ani in Alabama began to have what she called Bigfoot encounters at her isolated country home. Later, the creature told her that he was an Ancient One. His face is human and Ani has seen his mate with one child. Once she saw him dematerialize.

They became friends and, when he stood a couple of feet in front of her, Ani was going to hug him. Just as he opened his arms to hug her, two of her friends walked out from behind her house looking for her, and screamed, so the creature turned and fled. These encounters are profound and demonstrate the humanness in how the giants interact in a benign way when a person reaches out to them in a loving way.

This is a sketch by Ani of the Ancient One with whom she occasionally met.

Comanche, my Black Labrador would often literally telepath and "talk" to me, read my mind, and act like a person, which frequently freaked people out. He was evolved. When he passed away at age 13, it was devastating. I was blessed to have such a complex relationship with an animal. It's difficult to relate to other researchers after having so many powerful personal psychic experiences.

The day Comanche died, I took two boxes of Comanche's dog treats over to a friend's house to give Tara for her two small dogs. Comanche weighed 90 pounds, so I bought him the large milk bones. That evening while Tara's mother and sister were watching television in the living room, she walked in and gave each dog one of these bones. After putting the box of treats away, she returned to the living room to discover the two dogs fighting over one bone. The tiny 12-pound dogs could never have consumed a large milk bone in two minutes. Tara wondered what happened to the other bone! Her family looked everywhere but never found it. Shortly after, Tara saw a living apparition of Comanche in the house and realized he had "aported" his bone treat into the spirit dimension. Those were *his* bones and he wanted his share.

That same night while I was sleeping, my bed shook as it often had whenever Comanche had indicated to me that he needed to relieve himself outside. But he was dead! After he shook the bed with his paw the third time, I rolled over to see a living apparition of Comanche lying on his bed beside mine staring at me. I reached out to touch him but my hand went through his body, as his ghostly figure faded and disappeared. He came to me four times during a two week period.

Comanche was letting me know that he was still alive in the spirit dimension. Because we live in a dense material dimension, he could affect the physical bed from his etheric dimension but I could not affect his dimension. This is *exactly* the same phenomenon as with the interdimensional Sasquatch. While invisible or appearing as a living apparition, they can actually touch us or anything physical, but we can't touch or affect them. This is a vital concept to understand when interacting with the psychic Sasquatch, which I learned through personal experience and not by just concocting a theory. This is valuable first-hand data from the quantum realm. I feel that people should embrace the reality of these anomalies to see where they lead. Anomalies are the missing link everyone is looking for, but doesn't know it. It's really the New Science that has been there all along, that's an integral part of physics just waiting to be discovered by open-minded investigators who are intrepid enough to make a paradigm shift into this new and exciting reality.

The primary "tool" is mental telepathy. It's easier to learn than you might think. One first has to realize that the forest beings can read every feeling and thought. So think at them while visualizing the image of a Sasquatch. Hold onto that visualization and introduce yourself by telling them your name, then asking theirs. It takes patience and practice. Repeat the thought over and over again in your mind, then let go and relax. Be sure to shift from the intellect of your head/brain into your heart, with feelings of friendship, total acceptance, and unconditional love. I have successfully used this technique with dogs, cats, wild birds, and other wildlife.

Browse through Penelope Smith's web site at *www.animaltalk.net*. She is one of the top animal communicators on the planet. Take a workshop in mental telepathy, and it might change your life! Reading the book *How Animals Talk* (2005) by William J. Long can be helpful. *What the Animals Tell Me* (1977) is another excellent book, by animal psychic Beatrice Lydecker. *The Language of Animals* (2001) by Carol Gurney is superb. Then there is Daum Baumann Brunk's *Animal Voices* (2002) for more wonderful adventures that will aid you is trying to connect telepathically with the hyper-dimensional forest giants. Although the Bigfoot are not animals, the telepathic principle is virtually the same as used for animals. Remember to embrace the old adage "practice makes perfect." Everyone is different. One must start being consciously aware of every nuance, thought, feeling, and action. Many people are incognizant of a psychic Sasquatch, because they operate according to beliefs rather than by feeling what is really there.

On Planet Earth, *Homo sapiens* and Sasquatch-types worldwide *are* the exotic species. In 1985, when I asked a Sasquatch if they were the "missing link" that science talks about, he chuckled and said, "You people are the missing link; you don't know where you came from or where you're going!"

These various races of forest beings are our cohorts, our neighbors on Planet Earth. We would discover that we have more in common with them on many levels if Earthlings would only drop their prejudice and attempt to communicate and exchange ideas like a modern, "civilized" society should.

CHAPTER 5

THE VORTEX PHENOMENON: UNUSUAL CRYPTIDS

Buckminster Fuller said, "If you want to understand the human condition, you must first understand the universe."

Are there really doorways to parallel worlds? Are there different types of portals, which are opened by energy vortices—each leading to another "society" completely foreign to anything we can imagine? Personally, I know there are genuine portals that enter into a space and time inhabited by living beings.

I first read about an opening in our three-dimensional space in a book written by a journalist, while I was in Ethiopia, East Africa, on a 30-day safari in October 1967. It was a 3,000-mile trip through an ancient land from the lofty Abyssinian Highlands to the scorching heat of the Danakil Desert.

Each chapter delved into sensational topics that few, if any, bona fide scientist had ever explored. One chapter told of a parent who was awakened in the middle of the night to a commotion in his young son's bedroom. The boy had recently been given a puppy, which slept in his bedroom at night. When the father entered the room, the puppy had stopped barking and his son was calling his name from under the bed. Upon poking his head close to the floor, he was horrified to see just the two legs of his son as he wiggled further into an unseen hole. Less and less of the boy was observed until he nearly vanished into nowhere! The quick-thinking father frantically crawled under the bed, grabbed the child by the ankles and pulled him out of the invisible "doorway," until his son was safely in view. While being retrieved, the boy had a grip on the dog's leg and the animal once again appeared back in our three-dimensional world. The child told his dad that the dog had barked and jumped off the bed to inspect something under the bed. So the boy had followed. I vividly remember the story and felt relieved that the boy was rescued. Later, back in Asmara. I attempted to discuss this anomalous event with a nurse, but was told the subject was anti-intellectual and that I was wasting my time reading such trash.

Many years later, the Ancient Ones told me that these portals are on a "ley line" and the nucleus of a vortex moves up and down the line, based on astronomical movements and changing seasons. Some lines are longer than others. Portals differ from each other—some have healing energy, others are extremely dangerous to get near—but all open and close at certain times. Since a vortex can move through solid matter, that would explain it "appearing" at that unusual hour in the boy's bedroom. If the "door" closes when a person has accidentally stepped through, then that person is trapped forever and at the mercy of whatever living beings and animals exist in that realm. Scary.

I recall visiting the fascinating "Oregon Vortex" in Gold Hill in October 1987. A young guide escorted me, along with a dozen visitors, on an informal tour through the grounds, while giving us the history and letting the group experience its unexplained peculiarities. When we arrived at the old prospector's cabin, it was sitting lopsidedly on a very steep hill in the woods whence it had slid after a torrential storm in the early 1900s. She impressed us by holding a broom upright and letting it go, whereupon it stood upright on its own. Magic? No, the theory I had heard was that it is due to some kind of an electromagnetic force that apparently stems from mineral deposits deep in the Earth, that hold the energy in that specific area.

Then the woman told the group that the main vortex was 50 feet up the hill. This was incorrect. Being psychically sensitive myself, I could feel a strong energy vortex in the corner of the cabin. So after the 45-minute tour, I asked to talk with the owner so that I could tell her of the tour guide's erroneous statement. The lady-owner said that the guide was right: the vortex was up on the hill, and she claimed to be able to prove it.

As we left the kiosk, the owner picked up a set of "L-rods" that dowsers use to locate people, animals, minerals, water, and electromagnetic fields of energy. Though I am a Master Dowser who has done this work professionally, I said nothing. We walked up to the open cabin doorway, where she instructed me to watch carefully—predicting that the rods would point to the vortex above the cabin on the hill. She became befuddled when the wire rods pointed inside the cabin instead. Three times she tried, but got the same results. When she stepped inside the miner's shack, the rods instantly pointed to the corner that I had previously indicated to the guide.

I told her about the motion of a vortex nucleus along a "ley line." In the spring, when she checks it, the vortex will probably be 50 feet up the hill, slowly working its way down over the summer until it ends in the corner of the shack by autumn. This is really physics, not paranormal activity.

Dowsing is a real skill that is extremely useful when trying to locate a vortex in relation to the forest giants. However, even if a person finds a vortex, the hairy folks may not use it as a portal for weeks. Or, like the Oregon Vortex, it may be an entirely different type, that is not a doorway to anywhere. A few vortices (or vortexes— either is correct) have "healing" energy, and sitting on one can give some people relief from minor ailments.

On the other hand, my Ojibway Indian mentor told me in 1981 that she and a dozen other witnesses had seen an elder be incinerated while attempting to enter a vortex. This was around the early 1930s. Grandmother Kee said that the traditional medicine people of old knew the secret of how to step through this portal. They used to do it on a regular basis before the coming of the White man. She said there were

Starpeople living there who had helped her people in past ages. She told me that she had twice seen an ET while she was picking herbs near the "door." She also told of watching spaceships hover over the trees directly above the vortex!

People who have accidentally walked into a portal have reported almost a "void"—no sound, no wind, no sun, just a grayness with light—or even strange houses and unfamiliar landscapes. No one was dreaming, drinking alcohol, or hallucinating in any way. One man had his dog with him and the animal acted unusually—cautious, fretful, and reserved. When the hiker went back exactly the way he had come, without having been conscious of any "doorway," he was suddenly back on familiar turf. Often there is a time anomaly as well.

Brad Steiger, prolific author of the paranormal, collected information in his book *The Awful Thing in the Attic* (1995) that you would only expect in a science-fiction thriller; nevertheless, these phenomena are real, albeit unexplained. Steiger quoted a man who had had a few tantalizing encounters with interdimensional portals. The experiencer said:

> Some of these doors to other dimensions open like an elevator door with no elevator there to step into. Others open into a land of no life. Some take you back into the past, and some take you into the future on this world. Then there are doors that open into chambers that send the body to a distant star.
>
> This world we know as Earth is not the only world inhabited by people like us. We must keep our minds open wide.[1]

This same person said that at times he could peer into another dimension through what looked like a big window, yet he was unable to enter it. He claimed he could see real people moving around.

In July 2007 I was invited to meet a certain couple who lived in the Pacific Northwest. We had corresponded and occasionally spoken on the telephone over a period of about two years, so I looked forward to finally meeting them in person. They had a wilderness cabin and kept to themselves. I stayed with the couple for a week to get to know them better. During that time Jake, the husband, told me how a Sasquatch had walked up to him one morning while he was working on his porch. The creature came within seven feet of the man. It turned and smiled, then ambled around the house. The guy quickly looked around the corner of the building expecting to see its backside, but the giant had vanished!

One evening Jake spoke to me about a strange event that had happened in his front yard. A huge window—a "time portal"—had opened. As he watched, a street scene from around the 1890s came into view directly in front of him. Women wearing long dresses promenaded arm in arm with their suitors or husbands. Horse and carriages passed by along the cobblestone street. Jake yelled loudly to get someone's attention, since the people were alive, vivid, and strolling by only a few feet away, but no one noticed him. Jokingly, I said he should have tossed his cell phone through. But, he did not laugh, only looked at the place where it had occurred with great perplexity. In a short time the portal had closed into obscurity.

How credible does a witness need to be when gathering precious anecdotal information? In my estimation, Jake was an honorable man and detailed witness. "Let's not throw away the baby with the bath water," as astronomer Dr. J. Allen Hynek once told me during a meeting at which we had speaking of interdimensionalism and UFOs. He felt that anecdotal evidence was valuable.

Two days later after dinner, Jake and his wife took me for a walk to a small field—a clearing adjacent to their woodsy home. There were more surprises in store. They explained that one evening a few years before, the couple had strolled into this very clearing and stood talking. His wife began to notice what she thought was a swarm of large dragonflies that seemed to appear out of nowhere, approximately 20 to 25 feet away. The duo kept asking each other what these strange things were. Within a minute, one of the flying "insects" approached them and hovered at eye level in front of their faces. The beings were living, breathing *fairies*! They were flabbergasted. From what my friends could gather, the fairies had come through a portal from another dimension. They quickly concluded that they were actually seeing a real phenomenon and not just childhood folklore. Both people swore it was true.

Webster's dictionary defines the word *fairy* as "a tiny imaginary being in human form, depicted as mischievous, clever, and having magical powers." Omitting the word "tiny," one would think Webster was describing a psychic Sasquatch. Why? The common denominator here is that they are both interdimensional inhabitants of other-worldly realms, visiting our three dimensional physicality whenever it suits their purpose.

A very sophisticated and serious-minded friend named Kenneth Hunt (now deceased) was a frequent visitor to other dimensions. We spent many hours and days discussing these matters at length. Yet, Ken's labyrinthine thoughts and concepts were intellectually a strain for me to follow. He was brilliant, yet as soft-spoken and humble a man as you could ever find. During one of his dissertations, he said that the universe all around us is literally teeming with life, with every type of humanoid imaginable, including gnomes and fairies. Ken insisted that he was correct, in spite of my consternation and reluctance to believe at the time. This was in September 1986. After all these years of seeking the truth, I see that he was right all along.

There is an enormous array of uncanny fanciful cryptids being reported all the time, only to be dismissed by many as nonsense. So the real secret to it all appears to be the *portal*—the vortex phenomenon!

Portals may also be doors to other universes, intersecting at various points in the time-space continuum. There is substantial documentation going back to the 19th century of people literally vanishing in front of family and friends, never to be seen again. This happened in July 1854 to a Selma, Alabama, farmer, who evaporated into the ether while his wife and child witnessed his sudden disappearance!

Then in the mid-1880s a farmer outside of Gallatin, Tennessee, was walking across his field, as his wife stood in the yard to greet two visitors who were approaching in a horse and carriage. To everyone's horror, they observed the farmer suddenly vanish before their eyes. The people ran over to see if he had actually fallen into a hole, but there was none. He had unknowingly walked into an open vortex leading to another place or time.

In 1986, I read an interesting book about some new ideas and concepts in physics and parapsychology. Each chapter was written by a different author—a scholar who was erudite on certain subjects, with a new way of viewing science. There was one physicist who really impressed me with his ideas on interdimensionality. Then, in March 1987 I accompanied a friend who was driving to Reno from Oregon. When we stopped to stretch in Maryville, California, he visited an antique shop while I crossed the street to peruse in a dingy old second-hand bookstore. I immediately asked if there were any books on Bigfoot. The clerk said *no*. We started talking about the subject, and then I inadvertently got carried away and said that the creatures were interdimensional. I felt a bit embarrassed at first, not wanting to sound frivolous. The clerk immediately responded, "That sounds right. You should meet my boss, because a few years back when he was skiing down Mount Lassen, he skied right into a vortex and ended up in another dimension! He had to quickly dive back through so he wouldn't get trapped there."

I was delighted. How coincidental! How interesting that such events and phenomena are indeed happening around us. I asked his name so I could arrange an interview to glean more details about his observations while on the other side. I was stunned to discover that this man was the very physicist who had written the chapter about new frontiers in science. However, my repeated calls to him were never answered, as he was traveling frequently at the time, so I stopped phoning him. It would have been an excellent opportunity to get details of the mysterious other-dimensional milieu.

In September 2004 I met a sweet lady named Terry with an infectious smile as I was inspecting some wilderness property in Washington State that she was overseeing. Less than ten minutes into our hike, three Sasquatch astrally appeared and stood between us. By her behavior and body language, I knew that she knew they were there. So I told her they were there, and she replied, "Yes, but how did you know?" Terry was very psychic and well read, and owned a horse ranch in the deep forest near the Canadian border. She was also very conservation conscious when working her ranch. I told her that I knew where there was a powerful healing vortex on the property we were looking at, and to my surprise she pointed to the exact place. We burst out laughing! This is how intuitive people validate one another, which in a sense, is using the scientific method to confirm the hypothesis!

As the day went on, she opened up to me by sharing a little secret about her ranch. She had discovered a dangerous portal in her field that "felt" strange, so she always walked way around it. Being an animal lover, Terry had had three dogs in her life. One day her retriever walked into the invisible but higher energy portal and vanished before her eyes! She never saw him again. Her horses were using that pasture, so she decided to put a protective wire fence around the hazardous area. The day she brought in the fencing material, one of her horses was perilously near the portal. When part of his body began to disappear, she quickly whistled for him, which caused the horse to turn and gallop over to where she was standing! Everything Terry related about portals, healing zones, Sasquatch on her property, and so on was in full agreement with what I have learned over the years.

In the spring of 2009 I interviewed a man from Ohio, who told me about an experience he had had while mushroom picking in April 2001. George became very emotional while sharing his story. He began by telling me that he's a rather large man and had never been afraid while in the forest. Most of his life was spent in the woods and he enjoyed it. Both he, and at a later time, his wife had encounters with a Sasquatch that he considered something special.

On one occasion, George was mushroom picking with a good friend. As they hiked back toward where the truck was parked on an old dirt road, he decided to check out one more area just before they left to go home. Since his eyes had been focused toward the ground as he was walking, George was startled when he looked around and discovered that he was somehow in an entirely different landscape that was unrecognizable to him. He had become lost!

As he ambled along through the thicket, George began to notice a strangeness about the forest. Where did the prehistoric-looking six-and-a-half foot tall ferns come from? The exotic botanicals amazed him. Then there were these unusual extra-long flowers that he had never seen before. Out came his camera to take a photo but was chagrined to learn the camera wouldn't work. He also noticed that a heavy, misty atmosphere prevailed. He became scared! Everything was unfamiliar. The trees, the flowers, and the sound of jungle-type birds were foreign to him.

George came to a section of the forest where huge leaves were arranged in such a way that he was convinced that a real dinosaur was going to leap out at him at any moment. Everywhere he walked was spooky. He was convinced that he was in a real Jurassic Park and this caused him to become panicky. He wanted out!

Finally, George stopped, sat on a log, and got out his water jug for a drink. As he anxiously gazed around the primeval forest, to amuse himself he yelled out in desperation, "Hey, Bigfoot, I'm lost. Come and help me!" After placing the water jug in his pack, George felt as if he was being watched. Looking up, there nearly 70 feet away, standing by a large tree he saw a ten-foot-tall Sasquatch. The giant simply lifted his arm pointed the direction in which George should walk, then dematerialized. He gingerly headed in that direction. Soon, he noticed the sky was once again blue and the sun was shining. Just as the "real" world was unveiled, George heard a "swishing" sound, like a sucking of air. He told me he felt relieved to be out of the "twilight zone." Looking down on the ground in front of him, he saw a set of Bigfoot tracks. Then he heard his buddy calling him. As George approached the vehicle, his friend asked where he had been for the last two hours? The interdimensional interloper was sure that he had been gone only for a few minutes!

There are two documented accounts of human-looking people crossing over into *our* three-dimensional world and not returning to where they came from. I can only find one of the cases on file. In August 1887 near Banjos, Spain, two green-skinned children walked out of a small cave into a very alien world—at least our world was foreign to them. The boy and girl were confused as to how they got there. It must have been frightening for them to experience having been instantly transported to a strange world. The youngsters wore clothing made of shiny fabric unlike anything the local farmers had ever seen. A family immediately adopted the green-skinned children, but the boy died soon after. In time the girl learned Spanish and was able to communicate to her new family that she had come from a sunless land and that a strange whirlwind (vortices are spirals of energy that one might interpret as a "whirlwind") came and suddenly the two children found themselves in a cave somewhere in Spain. In time the girl's skin became less green until it completely faded, perhaps from a different diet and a sunny climate. Eventually she too died young.

I recall that a second green-children story was documented by a vicar in the 11th century in Woolpit, England. They were also deposited in a cave (probably through a portal) and again mentioned a sunless land as their origin. The boy died, but the girl lived and learned English, and her skin color turned "normal" over time. When she reached adulthood, she married and led a typical British life. In terms of genetics and DNA, she must have had 46 chromosomes like other *Homo sapiens* in our world.

Then how did their race evolve? How would Darwinists, Creationists, or Interventionists apply their cherished beliefs to this precarious position in anthropology? I include these many portal accounts because of their direct relationship to the Sasquatch people, ETs, and rather astonishing out-of-place cryptids.

There are several alien-animal types being reported with a degree of consistency that parallels the Sasquatch phenomenon. As an eclectic researcher, I draw knowledge from other disciplines whenever I begin to see similarities to the Sasquatch. For example, in the rural community of Danville, New Hampshire, a few years ago (posted on *cryptomundo.com* on January 8, 2008), there were at least ten reports of a giant monkey, and the local fire chief was one of the witnesses! It was said to be eight feet long, which included its "tail."

One of my alma maters is the University of New Hampshire—Durham, and during the summer I lived in a wild, forested region while working for the state forestry department. My degree was in Soil and Water Conservation. I hiked in numerous isolated areas every day. Often, I would question foresters, survey crews, wildlife biologists, farmers, hunters, and game wardens about what animals inhabited certain areas. No one, in 1962, said anything about an eight-foot monkey. My point is: Where could it have come from? We have credible witnesses that say the "animal" had to come from somewhere unique. In some cases, the worn out statement that a pet owner released it might be true, but certainly not in all!

It is my contention that a whole array of "things that go bump in the night" are coming and going through a vortex! Whether the giant monkey was a variant of a Sasquatch-type or something altogether different, no one can be certain. But I am completely convinced that the vortex phenomenon is the answer. In a quantum sense, there must be numerous parallel worlds up against or on top of the physical plane in which we live.

By age sixteen, I had read and studied every book in the local library in zoology and I have continued to stay updated over the years. I feel that cryptozoologists may be confusing interdimensional cryptids who are vortex users from an exotic dimension with rare or undiscovered zoological animals. The vortex users visit here to exploit our world by either preying on animals or eating vegetation. It's akin to a scuba diver exploiting an underwater world to get fish and mollusks, then returning to the air-breathing realm.

The chupacabra or "goat sucker" may easily be another example of a vortex user. The chupacabra has been plaguing many regions in Puerto Rico, Mexico, Texas, and Arizona. And there is a place in the mountains of northeastern Tennessee where I interviewed a woman who said the chupacabra have been seen there for decades. This may be a commonly accepted creature in that region of Tennessee.

In Linda S. Godfrey's 2006 book *Hunting the American Werewolf*, she has researched and collected demographic data showing that Wisconsin has more reports than other state of man-like, hairy beings with a wolf's head and paws, as large as a deer, usually bipedal, but at times running quadrupedally! There are also sightings from the upper Midwestern states, and from New York to the southeastern states as well. Some wolf-men were reported to be six-and-a-half to seven feet tall when ambling on two legs. I had heard of werewolf-types being encountered in different states, but wrongly assumed that it was merely a Sasquatch being misreported. After reading Ms. Godfrey's book, it was obvious that there was an entirely different phenomenon and, once again, the most likely source of entry into our world is a wormhole.[2]

For eight years I lived in Wisconsin and interviewed people who encountered Bigfoot and UFOs, but never knew how prevalent "wolf-man" sightings were. I do not believe for a moment that people are transforming into werewolves during the full moon. On the television show, "Monster Quest," three witnesses observed three wolf-men during daylight hours drinking from a stream. They immediately fled the area.

Just when I think I'm close to having a better understanding of the Bigfoot phenomenon, a witness informs me of something queer and extraordinary. One day I visited a woodcutter friend who supplied me with cord wood. He was standing in his yard talking with a friend whom I did not recognize. When he introduced me to Jerry, he added, "Kewaunee is the guy I told you about that wrote a book on Bigfoot!" Then he proceeded to coach Jerry along by telling me about Jerry's Sasquatch encounter. But Jerry was reserved and a bit distrustful, and never said a word.

A month later I stopped again and Jerry was helping him in his shop, fixing a truck. We chatted for an hour and Jerry surprised me by saying that he would like to buy a copy of my book. Then he told me that in 2006 while working over on the Olympic Peninsula, he and a friend had driven down a logging road to check some trees. They were sitting in the truck, smoking cigarettes, and just as he was about to step out of the vehicle, Jerry was alarmed to see what appeared to be five Bigfoot creatures some 25 to 30 feet away, standing and looking at them. One of the beings, who was six feet in height, walked up to the truck, ducked down, then slowly rose up, pressing its face against the window to see inside. This freaked them out!

Jerry kept stopping in the middle of his story about the encounter to test my reaction, because he didn't want to be made fun of or get into an argument about its authenticity. He continued by stating that the group of creatures was milling around and when he stared at them to get a better look, he noticed one that was very different from all the rest. It had a wolf-like face with an extremely long neck that made it look awkward in appearance compared to the others. Then Jerry told me it resembled Anubis, the ancient Egyptian god of the Underworld, except for the odd long neck! He stared at me in a serious manner and said, "Kewaunee, I'm telling you the truth!" Also, he would not tell me specifically where the encounter occurred, because he did not want anyone disturbing the area.

Anubis is depicted in hieroglyphics as a being with black skin with the head of a wolf, but with a normal human body and neck. Was the Anubis-type being a genetically engineered humanoid creature dropped off by ETs? Had the group just stepped through a portal, because the beings had not been there when they stopped the truck? It's very baffling! After Jerry left, my woodcutter friend said, "He's a hell of a nice guy, and he's not a liar either." Jerry still sticks to his story. At times it's mind-boggling to sift through such data, yet it is potentially valuable information. I have included Jerry's account with hopes that a reader might have had a similar sighting and would contact me and share it.

Could it be that wolf-men, like the Sasquatch people, are humanoids who have tremendous reasoning abilities, use mental telepathy, and have families on the "other side"? I feel it would be useful for Bigfoot researchers to read Godfrey's book to better understand other parallel phenomena to compare against Sasquatch behavior. The association with *portals* amongst these alien-animal phenomena is the factor that stands out the most. To truly understand the phenomenon and solve the Sasquatch mystery, it is important to take a long hard look at other overlapping phenomena—no matter how extraordinary they appear on the surface. Thus far, the research into the nature of Sasquatch people has been very superficial, because the majority of people are looking for a non-human primate!

On April 1, 2008, a First Nation woman contacted me from eastern Ontario to say, "I have seen not just one, but a family of Sasquatch." The woman was distraught about the visitors, which her family continued to see. On separate occasions a Sasquatch telepathed to her and later her son. She also was seeing strange-looking black creatures that disturbed her. Later she said, "Then the wormhole thingy had a little man coming out of it." The wormhole or vortex was located in the forest, a short distance behind her house, where it's heavily wooded. She was so upset that she broke down and cried.

The association is very clear, based on consistent patterns that have been reported over the years: Sasquatch, smaller humanoids, alien animals, mental telepathy, and a vortex—all in one geographic place. I have spent two to three hours at times counseling people, empathizing, sharing, educating, and helping them place the experience in an acceptable perspective. Often they have multiple encounters, so I teach them telepathy and the proper thought patterns and demeanor, so their actions create a successful response. The Sasquatch people have asked me to do this, as part of my spiritual job here on the physical plane.

The Mutual UFO Network database reported that on December 31, 2009, an Oklahoma witness claims to have observed a wormhole opening up in the sky next to the rising full moon, and a UFO coming out of it! Experiencers need support from someone who understands these phenomena. They should be able to put the anomalous event in a comfortable perspective without being ridiculed. In this case MUFON provided this service.

These unusual cryptids *must* be coming through portals, because Sasquatch, Ancient Ones, ETs, and black cats are well known to be vortex users. The Sasquatch and Ancient Ones make matter-of-fact reference to black cats, and one Ancient One had a large red wolf with him in Texas and later a black cat!

In October 2004 I was amazed to see a large, pure-black cat standing in a field near a road. It appeared to be around 60 pounds, about 4 to 4½ feet long, with a tail like a cougar, except it was slightly bushy as it arced upward. When it turned its head, I was surprised to see "tufted ears" and slanted eyes. The animal did not frighten me. I instantly knew it was of no known species.

Later a farmer showed me a photograph of a black cat hanging out in a Bigfoot area. And while weeding his garden, he had seen the pet wolf that the Ancient Ones had. This man had a college degree, was well read and woods-wise, yet he claimed that he and his wife had seen a "wolverine" crash into their fence one day while they were working around the garden. They were in such disbelief that they went on the Internet and found a photograph. Both agreed that that was what they had seen in the farmyard. If these two witnesses are correct, what was a northern-dwelling wolverine doing in Texas? Where did it come from? A vortex? We can only speculate.

While we might presume that local people would know what's in their area, I've found that frequently they really don't. While living in Ethiopia, I was told that aardwolves were extinct. Yet a few months later I met a man who had four pups in a basket that were clearly aardwolves. In the Awash region I saw kudu, baboons, gazelle, warthogs, and lots of fascinating African wildlife. Later at a small hotel in the village of Awash, I was told there are no more cheetahs in the Awash because poachers had killed them all years ago. That's one animal that I wanted to see. The very next morning from our Volkswagen bus, some 30 minutes from the hotel, we saw a beautiful cheetah standing majestically beside the road. My companion and I both had cameras around our necks, but were so enamored by such a rare sight that neither of us thought to take a picture!

I have been collecting information about large cats, particularly black panthers. There is a plethora of information on the Internet, and a few reports go back into the 1800s. The literature from zoologists say melanism in cougars is rare, but is more common in jaguars. I feel that public administrations in state wildlife services often rationalize that a black panther sighting is a melanized cougar or jaguar, because there is no other publicly acceptable explanation. But if it were a "normal" melanistic jaguar, then where are the reports of children or adults being attacked and eaten? Humans are easy prey. To just say that someone released a pet panther would mean that there sure are a lot of black panther owners in the USA, United Kingdom, and Australia, who are inexplicably releasing them. How could that be? There are too many "pet panther theories" to account for what appears to be a global phenomenon. Asia, South America, and Africa are not included, because these are areas where one would expect an occasional rare black panther.

In the *Press Democrat* on May 3, 2008, out of Santa Rosa, California, an article states that a man photographed a black cougar. Another black panther was spotted in Anderson, South Carolina, as posted on *cryptomundo.com* on March 30, 2007. KSBY6 Action News covered a story near Avila Beach, California, where an enormous black panther was seen. It aired on April 10, 2007. There are several more reports from California on file. The United Kingdom has been literally plagued with large cat sightings that are too numerous to list—not to mention a black panther killing a sheep in Northern Ireland, no less. This was released on March 21, 2008, by the BBC, and reports go back to 2003. Where are all these phantom animals coming from? Huge black panthers are being seen consistently every year in these regions.

The Lake Placid News stated that Dan Plumley and his wife spotted a black "leopard" in Keene, New York, on April 24, 2008. The state of Louisiana was not to be outdone, with one black panther killed and a live one photographed. My feeling is that the sheriff and/or a wildlife department don't want to excite the townspeople, so they may have buried it without biological analysis or, if it was autopsied, they are not sharing the results with the public. US Forest Service employee Terrance Fletcher in Georgia has had no doubt about the existence of black panthers since the seven foot animal charged him in January of 2007. Fletcher leaped into the cold Chattooga River to escape the cat. So there are rare incidents where nonfatal attacks by these cats have occurred.

A Newton County deputy in Neosho, Missouri, "shot and killed a large black cat of uncertain species," according to the *Joplin Globe* on May 22, 2008. The black panther charged the deputy sheriff, so he fired on it. In the newspaper article, the animal was described as a leopard or jaguar. It's odd, but I can't seem to find out if a zoologist properly identified the animal. Mississippi has had its share of black panther sightings, as have Pennsylvania and Ontario, Canada. In September 2007 Spain had several sightings of a black panther and, like a majority of cases, the investigators were unable to hunt it down. Can we assume from nearly all of these reports that black panthers are "still" roaming the land? Why is it that the encounters wane and then re-emerge years later? The large cats have to eat and yet there have been no more reports of livestock killed. There is an ecological piece of the puzzle missing in this scenario.

Now for a shocker! *Cryptozoology.com* had a February 22, 2007, posting about black panthers in Hawaii! How did that happen? These large cats must be Olympic swimmers! I found an interesting message on this website from "cryptojunkie" saying,

> It was like 3–4 years ago that a black panther was sighted here in Hawaii. There were links posted on here about it. It seems the panther has resurfaced and was on the Hawaii News again tonight. Can anyone provide me the links to the old articles?

Someone was kind enough to e-mail cryptojunkie the June 17, 2003, article from *HonoluluAdvertiser.com*. The island of Maui was where nine sightings occurred over a six-month period of a 100-pound, dark brown cat with a long tail. Apparently the reports were numerous to a point that Arizona biologist Stan Cunningham was asked to fly to Maui to assist the state wildlife biologists in an attempt to snare the animal. From the sightings, expert Cunningham had estimated that the cat was a 150-pound or more leopard or jaguar. To date no black panther has been captured since 2003, as it seemed to simply "vanish"!

Cryptojunky's reply on February 22, 2007 was,

> It seems the cat has moved islands and now is on Kauai. This time they had pictures on the news that a tourist took. If I remember correctly they were unable to find the large cat even with bringing in experts to track it down.

There are very strict laws in Hawaii regarding entry of exotic animals, regulated by quarantine laws because of environmental impact. Fines are $25,000 or more. So why would a person risk sneaking a potentially dangerous animal into the islands? The "pet" explanation doesn't make sense.

In June 2005 I was on Kauai for ten days, camping in the jungle. My Sasquatch friends in the Pacific Northwest arranged for me to interact with the Sasquatch on Kauai, where there are twelve. Maui has six. Also, there are two active portals on both of these islands—the exact same islands on which the black panthers have been seen! It's no coincidence. Different species of cryptids, ETs, Sasquatch, and the Menehune (Hawaiian little hairy people) also use them. It is my hypothesis that these black panthers are very capable of finding and using a vortex for safety reasons and to return "home." If so, they must be also using wormholes in the UK, Australia, Ireland, Canada, Spain, and the continental United States, from which reports continue to come in. People think that these out-of-place animals could never have the intelligence or ability to "dimension hop" via certain energy vortices. But this consideration is of the utmost importance when evaluating the

Bigfoot phenomenon. It is too uncanny that a panther appears 200 miles away on another island without any "normal" means to get there.

The Sasquatch people acknowledged that there are portals on Kauai and Maui to which they have access, and that there are none (at least that they can use) on any other Hawaiian Island. Researchers say there are no Sasquatch in Hawaii—their reasoning being that none have been reported. One may ask: how did the Sasquatch originate on a group of isolated islands nearly three thousand miles from the North American mainland? My answer is: either through a vortex or most likely dropped off from a spaceship by their friends the Starpeople, who also use the portals.

Along the same line, what about reports of people seeing African lions? In early 2000, a man in Idaho shot and killed an African lion in his backyard. Immediately, it was discovered that the cat had gotten loose from a menagerie not far from the man's house. No mystery to its origins here—***if*** that were the only case.

Beckley, West Virginia's *Register-Herald*'s October 25, 2007, edition announced that an African lion was seen hanging out by a hunting shanty for 40 minutes. The tail was four feet long, the witness said. An animal refuge was contacted, but there was no missing lion. Thus far, no other reports have come in from West Virginia. It just seemed to vanish. On January 26, 2008, a bus driver near England's Nottingham

This is the rugged interior of the island of Kauai where the Sasquatch live and where Jurassic Park was filmed.

Forest reported seeing a large, maned lion cross the road in front of him. The story was printed in the *Retford Trader and Guardian* newspaper. New Zealand was plagued by both lion *and* black panther sightings by credible people outside of Auckland in the Kaiwaka area. This was posted on Cryptomundo on July 3, 2008. And the El Paso County Sheriff's Department and the Colorado Division of Wildlife were in a conundrum when on July 22, 2008, an African lion was encountered east of Colorado Springs. A witness took a blurry photograph of a large, maned animal with a long tail, reported the *Colorado Springs Gazette*. The Colorado State Patrol, Cheyenne Mountain Zoo, US Department of Agriculture, and the police helicopters that assisted in the search found nothing—once again, the lion seemed to strangely disappear!

In December 1991 I interviewed a man for almost two hours after my lecture in Las Vegas. The man was tense and afraid that I would not believe him when he told me about some woodland he had acquired in northeast Massachusetts, close to the New Hampshire border. One day while hiking to survey the forest in the far corner of his property, a neighbor lady happened to be walking through the woods on the other side of a barbed-wire fence. He introduced himself and they stood and chatted for some time. The lady told him in a mysterious way that he had a portal on his land not far from where he was standing. Since he didn't believe that and was not interested in the subject matter, he merely dismissed what she had shared. A couple of months later, while hiking in that same area, he told me he was "scared out of my wits when I encountered a huge African lion as big as a pony!" The witness said the cat looked directly at him with intelligent, human-like eyes, then nonchalantly ambled off. It was how the animal looked at him that put him at ease. The man swore he was telling the truth.

Lastly, I interviewed a woman from northeast Pennsylvania who told me that in late spring 2006, while she was driving on a winding mountain road just as dusk was approaching, an African lion jumped over the hood of her car! She was driving slowly to negotiate the curves when the huge cat startled her. It was not reported to the police, because the witness did not want to be ridiculed. She perused the newspapers daily and watched the local news on television, but nothing was chronicled pertaining to an African lion. A side note on this report: While visiting Pennsylvania the last week of October 2005, I interviewed a man who claimed he had frequent visits from the Sasquatch people. I spent two days discussing the details with him. On the second day I located a vortex in a heavily wooded hillside above his small farm. He told me that he and other witnesses had seen the hairy folks coming from that hill. The woman who encountered the lion lives next door to this man, and it was at the base of the hill where it leaped over the front of her car. From the road it's less than a hundred yards up the hill to the portal! Coincidence? Maybe, but I look for patterns of behavior in what is being reported, and over the years my "vanishing cryptids theory" continues to point to the vortex phenomenon.

I find anecdotal evidence from seemingly credible and sincere witnesses valuable 99% of the time. So I empathize with experiencers who have had no outlet—no one with whom to share. I help them analyze something that was very real and traumatic to them, and out of the bizarre and outlandish usually comes a valuable piece of the puzzle.

There is also the giant snake phenomenon, with many encounters in the more northern climes. I am not speaking of constrictors that escape into the Florida Everglades. That is a serious problem in itself. I am referring to places in Oregon, Kentucky, and in the spring of 2009, in eastern Iowa. I interviewed a woman who innocently began having Sasquatch encounters, after which alien animals started showing up. The witness said a neighbor and family members could testify to her experiences. Because she had been badly ridiculed by friends and neighbors, the woman was most reluctant to open up to me. After reading my book, *The Psychic Sasquatch,* she felt more comfortable and decided to talk. I have met several witnesses who, similar to myself, have had multiple encounters with a variety of beings and cryptids. This woman is one of them. She is a person who has a certain personal, emotional, and spiritual qualities that attract these experiences. Lack of fear is one of the most important characteristics with these types of people.

The witness said it all began when a Sasquatch showed up on several occasions, and both she and her son began having mental telepathy with the man-creature. One evening she went outside and stood by the light that was shining from a large window in back of the house. Out of the dark stepped an extremely large black panther with small saber-tooth–like fangs in the upper part of its mouth. The animal appeared relaxed and docile. It sat down about eight feet away and casually looked at her with a friendly expression. "Were you afraid," I asked. She said, "Not at all." Eventually the cat stood up and walked into the night. What was it and where did this out-of-place, pony-sized felid come from?

This contactee lived close to a river and took her young boys fishing on occasion. After cutting the heads off the fish and gutting them, she would bury the remains in her garden. Then she discovered that the giant black panther was digging up the decomposing fish. The tracks left in the soil were as big as four packs of cigarettes placed together, she said. When the Department of Natural Resources came out at her request to evaluate the paw tracks, they estimated that the cat had to weigh around 700 pounds!

The Iowan lady said that only once did she feel intense fear: she stepped out on her back porch on a clear moonlit night and saw "at least fifty Sasquatch with red glowing eyes walking across the field." She indicated that there were families of them and they were easy to see in the moonlight. Apparently, they were migrating to another area. The fear she felt made her retreat immediately back into her house. I think that, because the giants in this tribe had their children and families together, the males clairsentiently sent her a message to leave, since they were caught in a vulnerable position.

A month later, while sitting in her car at the farm with the window down gathering her thoughts, a gigantic wolf-like canid walked up to her car door. Its huge black head was level with the open window. As the woman turned around to get a better look, the animal gently stuck its massive head into the car, licking her in the face before she could react. Again, the experiencer said she doesn't know why, but she had no fear. The black "wolf" then slowly walked away. Where on Mother Earth would a cryptid such as this come from?

Additionally, the witness saw a werewolf-type bipedal creature that she said the Sasquatch do not like. Then one day, as she was slowly driving a truck on a wood-road on the back side of the property, the woman stopped to remove a tree, twice as big around as a one-pound coffee can, that was blocking the road. Her thought was to pick up part of the "tree," pivot it around and to get it off the road. The witness was stunned when grabbing the "log," that it began to move. To her horror, it was a hulking snake—30 feet long or possibly longer! The head was concealed in the brush, as was its tail on the other side of the wood-road.

A few days later, the woman spoke with a neighbor who lived adjacent to her place, and the farmer told her that he had encountered a 30- to 35-foot snake coiled outside of his barn. This man was 6 feet,

4 inches tall and the serpent's head was the same height as it stared directly at him. There are a number of questions that she couldn't answer about her neighbor, such as: Did he run? Did he stand and watch it slither away? What color and markings did it have? Did it look menacing? Were there any farm animals missing?

This Iowa experience was similar to that of a woman in Tillamook, Oregon, who encountered a snake while blackberry picking. Later, a pack of black wolves was seen chasing a cow in the pasture; plus she had had Sasquatch visits and seen a glowing spaceship by a vortex about 80 yards from the house. The common element, again, is a portal.

Other animals and critters have been puzzling and scaring witnesses—like moth man, thunderbirds, even pterodactyls.

I interviewed a woman in Iowa who told me that her next door neighbor walked up to two six foot tall eagle-like birds standing on a river bank in the very northeastern part of the state. Jeremy Lynes, one of my best friends from Georgia, told me his ex-girlfriend was adamant

Experiencer Jeremy Lynes with his Russian Wolfhound "Peaches"

about encountering a pterodactyl flying overhead in North Carolina. She was so serious that she warned him not to ridicule her about the matter, insisting that it was not a case of misidentification. Why would people set themselves up to be poked fun at unless they had actually had an authentic encounter? Jeremy himself has had several Sasquatch encounters, including being touched by one. He is a very articulate, well-read, and aware person.

On the Pennsylvania website *www.stangordonufo.com* in January 2009 it was reported that a giant prehistoric bird-like creature was seen in February and May of 2008 in that state. In one of the accounts, there were several witnesses, since the flying creature glided low over cars on a busy road. "One witness noticed that the wingspan extended beyond the edges of the two lane highway." Now, that is one big unidentified flying object! Average citizens are reporting these raptors annually. They would have to have nesting areas if they were indeed from our world. Where are they? None have ever been reported.

In Alaska in 2002 a pilot and several passengers observed a gigantic eagle-like bird that glided next to their small plane for all to see. It was reported that the wingspan of the creature was equivalent to the wings of the plane. Was it a visitor that came out of an aerial portal?

Enormous flying creatures or thunderbirds have been reported to be the size of small aircraft. Sightings were reported in Oregon in 1954, Kentucky in 1966, West Virginia in 1967, and Massachusetts in 1995, to name a few. Pennsylvania has had over a dozen accounts since 1940.

Pterosaurs and different types of pterodactyls have been reported on numerous occasions around the United States, Latin America, Africa, and New Guinea. It's difficult to sort out zoologically. Some exotic species are most likely coming through a vortex, but others may be Earth-evolved cryptids that have not been "discovered" thus far. From my decades of research, this conclusion seems to be the case, especially based on my ongoing rapport with the Sasquatch people. The subject of portals is very common when conversing with them and with several other contactees. Often they do not go out of their way to elaborate on a subject, because they wish to protect their survival secrets from threatening humans.

People should not assume that if a being doesn't have some *technology* to locate portals, that these "dumb animals" couldn't find them. Lots of animals—crocodiles, sharks, dolphins, and so on—have different innate sensory "devices" to locate prey in murky waters. Likewise, we must not underestimate alien animals just because they look weird and out of place. The fact that we don't have a specimen of one of these anomalous visitors indicates that their perceptual skills surely outweigh ours.

Establishment of a portal affiliation is absolutely vital to understanding the Sasquatch and UFO enigma. For the Sasquatch, a vortex is a refuge—a safe zone to which to flee from Bigfoot hunters and other dangers. It's a sanctuary into which humans cannot follow, and is one of the primary reasons why these beings and other elusive cryptids can't be found!

One book with much insight is *Merging Dimensions: The Opening Portals of Sedona*. Coauthors Tom Dongo and Linda Bradshaw state:

> After ten years of almost full-time UFO and paranormal research here in Sedona, I have come to the conclusion that, along with the world-famous vortexes, there are also interdimensional portals in this area. These portals, or windows, seem to be entry and exit points into another dimension or universe or place that we, at this point, do not fully understand. This sort of thing isn't only happening in Sedona. Around the world there are a number of these anomalies developing or, perhaps better put, just being discovered. My estimate would be that there are probably hundreds of these portal-type anomalies spread around the globe.[3]

In their book numerous documented encounters of aliens, UFOs, Sasquatch, and out-of-place animals are discussed—and they are usually found near a portal.

A submitted article on *bigfootencounters.com*, entitled "The Diane Vaughan Story," dated May 17, 1989, was told to UFO researcher Peter Guttilla. The encounter took place in the vicinity of Altadena, California, at 5 p.m. in a mountainous region just above Pasadena. Ms. Vaughan was hiking on a trail in the foothills when she heard footsteps from above. She continued to follow a trail that led higher up a ridge. As she was trekking, a six-foot Sasquatch came walking down the trail toward her. In her excitement, Vaughan ran up the path toward the being, but it quickly ran off down a bushy ridge. She called, "It's okay, come back!" Moments later, "the Bigfoot reappeared on the path, just where he had left it. We stood looking at each other." In the report, Vaughan said it was then that she noticed a small rectangular area "filled with shimmering air." The shimmering was likened to heat rising in hundred degree weather.

> He put his arms into the shimmering area and they disappeared. He pulled out two brown arms. Their hands were clasped. They went back and forth a few times as if "he" was trying to persuade "her" to come through to this "side." Then he stopped and looked behind him and over toward me. And that's when I thought I should leave. I waved and he waved back.[4]

A man who I met while visiting a friend told me that the two men had decided to go for a walk in a meadow beside a wooded area behind the house. As they stood talking, a concentrated "wave of energy" became apparent on the other side of the field. Then a "wormhole" opened and a Sasquatch stepped out, looked over, gave them half a wave, then turned and walked back into the portal!

For years I subscribed to Ray Crowe's informative newsletter, *The Track Record*, to get an overview about what other people were experiencing. A person from Kentucky had submitted sightings that he and a neighbor encountered starting in 1975 near Spottsville. Then this person left the area, but returned to visit his old friend in February 2005. He was told by the old neighbor that Sasquatch sightings had continued around his place in the countryside off and on for years.

> Moreover, he said, that what he had seen with his own eyes went far beyond anything that he had ever dreamed possible. One day he was walking along a field and noticed a strange area that looked like "heat waves rising from a hot, summer road." The area was only a few yards wide and to either side everything looked normal. As he was watching, one of the creatures stepped out of this strange wavy area like stepping out of a doorway. One second nothing, and in the next, there it was looking right at him. It growled at him and, at the same time screamed inside his head to "leave me alone!" Then it turned around and took a step back into the strange looking "doorway" and disappeared. After that, he began watching the area from a distance using binoculars. In all, he claimed to have witnessed several different monsters using this doorway a total of three different times, always appearing or disappearing, seemingly, into thin air.[5]

On the night before I left Texas in October 2004, I was staying in a tent in a hardwood forest, and asked Haloti if she would ask the good ETs to take me through a portal. She said she would give them my request. Soon I fell asleep. In the morning I walked to my friend's farmhouse to have breakfast and get ready for the drive to the Dallas airport.

Before I continue, I want to explain that the farm family allows the Ancient Ones to come inside at night and take what they need, which is usually a large glass jar to make sun tea. So it is not unusual for them to find a feather or stone or crystal in the kitchen when they get up in the morning, because the beings leave "gifts of appreciation" behind.

Lying on top of my luggage was a note addressed to "Dream Man" (me) from Haloti. She wrote:

Dream Man

The Star People have
taken you IIIIIIIIIIII
cycles forward to
New Earth which
God has recreated

You have seen great
things and given much
healing lessons

Your memory of the
travel has been
erased but the knowledge
is inside you
Use it only for good
not for gain

A Star Man will travel
with you on metal bird
You will remember some
things when you
see him
God Bless You

I interpret Haloti's note to mean that I was taken as requested; that Starpeople took me into the future to 2016 after a possible global holocaust when ETs, Sasquatch, and Ancient Ones are working together to rebuild a new society aiding the human survivors. Apparently, I was shown some wonderful things that are unknown in the present world. Since I am a Master Herbalist, Haloti tells me I was taught some new healing techniques to help others that will "germinate" in my mind at the appropriate time, perhaps after 2012, similar to a post-hypnotic suggestion. Presently I have no knowledge of these lessons or of meeting anyone at my tent that night. There is no proof or indication that this event ever happened other than Haloti's written words. I believe her!

On the plane I checked everyone out to see if there was any clue of a Star man, but failed to notice anything unusual—that is, until I found my assigned seat. There was a quiet man sitting by the window and I had the aisle seat. No matter what I asked him or how cordial I was toward him, he never spoke one word all the way back to Seattle. He had a kind face, peaceful demeanor, and a gentle smile, but would not look at me. This is no proof of anything, yet in my deepest heart I want everything that Haloti wrote to be true. I am left only with a wish and nothing more. I look forward to the year 2016.

When I was on an expedition in the Ouachita National Forest in southeast Oklahoma, Pushoma, chief of the Ancient Ones there, claimed that he took me through a vortex for three days and showed me his world. Later, I asked Haloti what happened.

Haloti told me, "Time is not same at Sometime Place; you go for /// [three] days, come back to same time you left…hard to remember. God love you." So when he brought me back, it was the same Earth time, and I really don't remember. This annoyed me, and I wondered why he took me at all, when my whole intention was to experience another realm, then come back to report on it in detail. The Sometime Place is their reference to whatever is on the other side of a portal.

I am aware that they have impressive powers of the mind. But I was troubled over not being able to remember, and even wondered if I was being duped. When I asked Pushoma for a better explanation, he printed it out on paper and left it for me the next morning (a flair pen and a tablet of paper were always left out for them):

You world not same as
our world
You time not same our time
You space not same our space

A thin cloth hang between
There places that open and shut

There is no I or me
We are all one
Each one is part of the whole
What one does all do

You may see us as Brother Crow
or Brother Wolf

Listen to the stones
Speak to the trees

It's difficult to claim or accept the supposed fact that I was sent 12 years into the future by benevolent ETs and later spent three days in another dimension with the Ancient Ones. So I am very reserved about this and hope there will be another opportunity to address the matter with the giants in the future.

For now, the phenomenon of vortexes and the beings that use them will have to remain a mystery. The Sasquatch told me that the portals are used by them and cryptids. If and when one of the giants invites me again to visit its quantum realm, I would accept the invitation.

The Sasquatch people told interspecies communicator Kathleen Jones of Oregon, that the Loch Ness creature and similar aquatic animals world-wide—Thunderbirds, pterosaurs, black panthers, the lizard-men, chupacabra, et al.—are all beings from another dimension that frequently use "portals" to enter our physical world, and that's why no one has caught one! This is very important to understand. The creatures are animals with a higher consciousness that come from a parallel world. In their dimension there is no such thing as "strictly" a zoological specimen! The Sasquatch say we need to learn *equality, not* dominance with living things.

These experiences happen more frequently than the reader may know. Most experiencers are not talking. Now is the time to be open about all spiritual/psychic encounters, because the Sasquatch want that part of their reality to be known.

The Sasquatch people and ETs are not the only ones who think in terms of portals and other dimensions. A team of top world physicists is presently devising a means to explore this realm. Outside of Geneva, Switzerland, there is a 27-kilometer subterranean tunnel containing what is called the CERN Large Hadron Collider, which is a super-gigantic machine that scientists hope will open a door into a dimension outside the time-space continuum.[6] Sergio Bertolucci, Director for Research and Scientific Computing at CERN, said, "Out of this door might come something, or we might send something through it." Sounds like television's "Star Gate" to me, except it is presently happening on the threshold of our 21st century! Apparently this monstrosity of a machine can send high-speed hadron particles at velocities that nearly reach the speed of light—equivalent to 670,616,629 mph. These beams are

Rock stacking by the Sasquatch discovered one morning in front of the cabin in the Kiamichi Wilderness in Oklahoma

arranged in a way that produces colossal head-on collisions. The super-sophisticated "physics experiment" is apparently so violent when these streams crash that it's been predicted that a significant tear in the veil of time and space will occur, possibly creating a miniature black hole. Other highly speculative statements have been made as to what will occur, but no one really knows, since it's the first time ever that modern man has the equipment to tamper with nature on this level. I feel the Large Hadron Collider is really a dangerous toy built by out-of-control mad scientists who could interfere with the natural laws of our universe.

I don't know if a Sasquatch or ET can "build" or create a portal, or whether the portals are just there, naturally formed by cosmic events. My suspicion is they can produce a vortex similar to natural ones as long as certain geophysical conditions are right. Physicist Lisa Randall's book *Warped Passages* indicates that other entities that we could meet and learn from "may" exist in these extra-dimensions. My theory is that we Earthlings are the ones living in a lower, less advanced dimension than most or all interdimensional voyagers. It appears we are physical-dense, whereas higher-dimensional visitors are physical-etheric.

Dr. Bruce Lipton suggests that we can rearrange our DNA through our very thoughts, without the use of invasive technology. This means that, in time, earthly humans could potentially have control of sub-atomic particles via our minds. Thus we may have the ability to become interdimensional, like our big brothers, the Sasquatch and ETs.

In *Science Daily,* August 14, 2009, it was announced that Chinese physicists have developed the first electromagnetic gateway. The report was originally published in the *New Journal of Physics* by researchers from Hong Kong University of Science and Technology and Fudan University in Shanghai, and describes the concept of a tunable "gateway that can block electromagnetic waves, but that allow the passage of other entities," likening it to a "hidden portal" of the kind mentioned in science fiction.[7]

No doubt, an advanced race of Star beings that is a thousand years or more ahead of us, would consider this to be kindergarten technology compared to theirs. I was told that spaceships "dimension-hop" from wormhole to wormhole, allowing ETs to arrive here in a short few hours, Earth time. In linear space they may be 30 to 100 light years away, except that they take a short cut through a series of portals. There are terrestrial and extraterrestrial as well as aquatic portals on Planet

Earth that are being used all the time by advanced beings because it's so practical. This is the *real* secret of alien space travel when they visit Earth and other places throughout the vastness of outer space.

While on an expedition in Oregon, the author encountered two rare albino deer—not cryptids, yet seemingly out of place in the natural world.

CHAPTER 6

THE ELUSIVE EVIDENCE: REVAMPING THE EMPIRICAL MODEL

In 1895 explorer and famous Pacific Northwest mountaineer Major E. S. Ingraham claims he had telepathic communication for more than an hour with a psychic Sasquatch while climbing Mt. Rainier in Washington State. Ingraham later shared that profound experience in a book he authored, entitled *The Pacific Forest Reserve and Mt. Rainier—A Souvenir*. Why would a man with an impeccable reputation, who was well respected throughout the Pacific Northwest, make such a public statement unless it were true? Back one hundred and fifteen years ago, there were no Sasquatch researchers competing to find proof of their existence. Nobody cared or placed any importance in the phenomenon, because it wasn't even a phenomenon yet! In the chapter "The Old Man of the Crater" Major Ingraham writes about finding and entering a cave in the side of Mt. Rainier, then meeting an extraordinary man-being:

I quickly stepped within a recess in the wall of ice on my left and awaited developments. I had not long to wait, for almost immediately there came, now rolling, now making an attempt to crawl, a figure of strange and grotesque appearance, down the passage. It stopped within a few feet of me, writhing and floundering very much as a drowning man would do, when drawn from the water as he was about to sink for the last time. Its shape was nearer that of a human being than of any other animal. The crown of its head was pointed, with bristled hair pointing in every direction. The eyeballs were pointed too; and while they appeared dull and visionless at times, yet there was an occasional flash of light from the points, which increased in frequency and brilliancy as the owner began to revive. The nails of its fingers and toes were long and pointed and resembled polished steel more than hardened cuticle. I discovered that the palms of its hands and the soles of its feet were hard and callused. In fact the whole body, while human in shape, except the pointedness of the parts I have mentioned, seemed very different in character from that of the human species. There was nothing about the mysterious being, however, that would make it impossible that its ancestry of long ages ago might have been human beings like ourselves. Yet by living in different surroundings and under entirely different conditions many of its characteristics had changed.

By degrees this strange being began to revive. Gradually an electric glow covered the entire body, with light centers at the ends of those pointed nails, the eyes, and the top of the head. It seemed to accomplish its revivification by rubbing its hands vigorously together. As soon as it was able to stand, it began to rub its feet rapidly upon the floor of the cave. This increased the glow of its body and caused the light-centers to shine with increased brilliancy. It seemed to receive some vital fluid from the Earth that at once gave new vigor to its whole system. Involuntarily I imitated its actions and immediately found myself undergoing a very peculiar sensation. I seemed to be growing in accord with the strange being who then for the first time noticed my presence. He at once redoubled his former movements. He would rub his hands vigorously together and then quickly extend the points of his fingers in my direction when sparks of light would dart therefrom. Having become

> deeply interested in this strange exhibition, I went through the same manoeuvre with a similar result, although apparently in a much lesser degree. The effect was magical! I was becoming *en rapport* with the Old Man of the Crater! I could see a brilliant point of light gradually forming on the crown of his head. Feeling my own hair beginning to rise I removed my knit cap and felt my hair bristling upward to a common point. The light from his crown seemed to form an arch above and between us and *we were in communication*. There in the icy passage connecting the unknown interior of the Earth with the exterior, by means of a new medium, or rather an old medium newly applied, two intelligent beings of different races were enabled to communicate, imperfectly at first of course, with each other.
>
> For an hour I received impressions from the Old Man of the Crater. It is a strange story I got from him. While the time was comparatively short, yet what he told me, not by voice or look, but by a subtle agency not known or understood by me, would fill a volume of many pages. Finally expressing doubt at what he communicated, he commanded me to follow him. I had anticipated such a demand and was ready to resist it. So when he turned to descend, to the hot interior of the Earth as I verily believe, by a superhuman effort I broke the spell and hastened upward and back to my sleeping companions. This is no myth. The old man told me of his abode in the interior, of another race to which he belonged and the traditions of that race; of convulsions and changes on the Earth long, long ago; of the gradual contraction of a belt of matter around the Earth until it touched the surface hemming in many of the habitants and drowning the remainder, and of the survival of a single pair. All was shut out and the atmosphere became changed. Gradually this remaining pair was enabled to conform to the new order of things and became the parents of a race which, for the want of a better name, I will call *sub-rainians*. This Old Man of the Crater had wandered far from the abode of his race in his desire to explore. Far away from my home we had met, each out of his usual sphere.[1]

This is powerfully profound anecdotal evidence for the existence of a humanoid Sasquatch with psychic ability. Ingraham, I may note, did not refer to the Sasquatch as an animal but, from the telepathic conversation, deduced that it's an intelligent being of a human kind and labeled it as another *race*! What do conservative Bigfoot researchers and mainstream scientists do with such valuable information that is part of the historical records of their regions? Because it is anecdotal, they consider Major E. S. Ingraham's reality experience invalid in the eyes of empirical science! No doubt, Major Ingraham would be irate today if he knew that modern science was indirectly calling him a liar!

In the past we have exhibited our psychopathology in how we treated black slaves, First Nation peoples, and Asian immigrants. Now, with Sasquatch, let's start acting like a modern, 21st century civilization. It is vital to let go of the egotism and cultural pathology that are driving us toward chaos and extinction. Without tongue in cheek, I must say that the forest beings find our attitude toward them and the living environment most *abominable*. They say that Mother Earth is presently dying.

The major problem with the Sasquatch phenomenon in the empirical model is that science demands a live specimen, a cadaver, or bones from a decomposed body. Because the creature has hair (often mislabeled "fur," which only animals have) and is observed living in the forest, scientists make a subjective assumption that the beings are animals. It is customary for a scientist to dissect a specimen to properly analyze, categorize, and taxonomize an animal and gain knowledge of how it functions in nature. That is the process. No exceptions. Since medical and behavioral experiments are often carried out on either rats or nonhuman primates—that is, apes and monkeys—many crude experimentations are inhumane. Animals undergo torturous suffering at the hands of supposedly more evolved human experimenters. An oxymoron? Incidentally, the Sasquatch people are totally aware of this cruel and primitive practice. This is a primary reason why they won't let us acquire proof: the body is sacred to them.

Science mislabels the Sasquatch as an animal because it has full-body hair, in spite of it having a human anatomy, being bipedal, and having prehensile hands not claws. Since the idea of capturing a live specimen seems fruitless, because the giants are so cunning, then killing one is suggested as next best option to get the ultimate proof. Hunters, crafty woodsmen, and ex-military characters have continuously failed to bag one. We will never find bones of a Sasquatch or a cadaver in the forest, because they bury their dead just like us humans.

The second week of July 2010 a Sasquatch was hit and killed by a truck on Route 285 late at night outside of Bailey, Colorado. I spoke to a woman by telephone who had seen the corpse. She was a recent contactee who had previously known nothing about the phenomenon. On her way home from work, the Sasquatch telepathically told her to slow down so she wouldn't run into the elk-sized creature on the highway. They told the woman that their brother's name was Benerk. Shortly after she drove by the creature, the giants quickly retrieved their comrade's mangled body. The witness said she saw blood and hair, but never saw the face in the car headlights. The two Sasquatch who had been communicating with her were named Oberon and Fyla.

The giants do not want us to find any definitive evidence of their existence. Presently, in northeastern Oklahoma, groups of 30 red-necks or more "gang up" to sweep specific areas of the forest in an attempt to hunt down and shoot a Sasquatch. Shame on them! There is no justifiable reason for ever killing one of these nature people to acquire proof. The empirical model needs to change—to be more humanitarian when defining Earth's creatures in Universal Nature.

Though the Sasquatch people can transport themselves into other dimensions to evade stalkers, apparently they did so less in the past, because they were not aware how lethal rifles could be. They had never encountered guns until encountering the White man during the 1800s. A well-placed bullet can kill them. In Ohio in 1981 a farmer shot a Bigfoot and watched it fall to the ground, while a second creature came to its aid and helped his comrade back into the woods. Thereafter, the farmer saw several pairs of glowing red eyes staring out from the forest. All of his bullets hit only trees since the Sasquatch were peering out from the safety of another dimension! The precious physical proof was never procured.

In 1970 I studied the culture of South American Indians quite extensively. During my first trip to that region I covered a thousand miles by boat going down the Amazon River, then another 2,100 miles into Brazil's interior. Then in 1973, in conjunction with the Colombian Institute of Anthropology—Bogota, I conducted an ethnographic study living amongst the Tukuna Indians in Amazonia called the Upper Rio Loretayacu. There are still undiscovered tribes who hide from modern man. Not too long ago, in some areas of the Amazon Basin, there were bounties on Indians. Brazil sent in jungle commandos to "eliminate" the Indian problem! The practice of genocide has been effective throughout all of the Americas. Over 200 indigenous tribes have been exterminated in South America alone since the arrival of the rapacious White man. This was achieved through war, infectious diseases, removal from productive to nonproductive land, government handouts of non-nutritious foods that weakened the immune systems, and government "kidnapping" and displacement of Indian children. All of these horrific political practices are demoralizing and have spiritually disintegrated their native cultures.

That said, in which way will we solve the Sasquatch problem? In the Amazon some anthropologists and missionaries have brought gifts and kindness in an effort to appease potential tribal hostility. Pacification paved the way. Science will become balanced if scientists discover their *spirituality* on this issue by publicly announcing a strictly nonviolent approach to Sasquatch field research! Spirituality means having compassion and respect for all living things. Scientists could show that they are morally responsible as an establishment. Then they would earn the title "modern" in front of the word *science*.

Humans must humble themselves, put their egos aside, and begin to sincerely communicate with these forest people without any chicanery. Let's not have any more blemishes on us by repeating history. One has to understand that these beings are far more evolved and complex than any living "species" that science has yet encountered on planet Earth!

The only way scientists can be successful with these interdimensional beings is to change their understanding and recognize the validity of anecdotal data. They won't get bodies or body parts with these super-elusive people, so we must move on to the next level, which is to take anecdotal reports seriously as an integral part of empiricism. Why? Old-world thinking doesn't work, because of the ultra-perceptive nature of the giants. My years of Sasquatch research, during which they have been interacting with me, have taught me that the Sasquatch are in control, not us! The *only* research "tool" that genuinely works is replacing one's fear and violent tendencies with unconditional love, real compassion, and *trust*, because they know our every thought when we are in the field.

The next step is for field researchers to study and practice telepathic communication. Know that with this methodology, the giants will always be reading your mind and knowing your feelings. So, always remain *honest*—for there's a big shortage of that out there these days. This is the way for *Homo sapiens sapiens* to truly evolve in this discipline. Our technology has superseded our spiritual development, which is precisely why our cosmic ET neighbors are also avoiding us.

The psychic Sasquatch and other illusive anomalies cannot be adequately documented without the acceptance of anecdotal evidence; thus, the scientific paradigm needs to be revamped to include it. That way it better represents Universal Reality. Without this adjustment of the model, we are distorting and/or ignoring authentic evidence that can lead to empirical proof, although it is still up to the Sasquatch people to decide if and when to give any material proof.

Anecdotal evidence is much more reliable than some scientists claim. After documenting 187 accounts of psychic anomalies with the forest giants, I consider only two to have been hoaxes. Anecdotal reports are valid tools if the interviewer stays objective, sensitive, and thorough. When people encounter an aspect of a psychic Sasquatch, they *never* forget it for the rest of their lives. I have found that experiencers are often traumatized, fear ridicule, and have a cathartic need to talk to someone with a compassionate heart. Even though researcher Dr. Grover Krantz has always had my respect and was always cordial to me, I vehemently disagree with his statements that "for every small shred of proof, there are 100 hoaxes" and "most sightings and footprints are hoaxes." These statements are untrue and way out of proportion. Percipients deserve the highest respect for their courage in coming forward with "twilight zone" experiences.

Over the years I have collected physical evidence, some of it given to me by witnesses. Forensic scientist Dr. Kenneth Siegesmund is an expert with the electron-microscope who has analyzed my field samples for 30 years. Dr. Siegesmund is professor emeritus of neuro-anatomy at the Medical College of Wisconsin, where I also once lectured. I have presented track casts, hair, feces, and other physical evidence for his analysis. One plaster cast had dermal ridges. Also, I have recorded vocalizations, photographs, and video footage from other experiencers. No academician has ever shown any interest in the physical evidence I obtained. Whatever happened to those doughty, inquisitive field adventurers that delighted in exploring new mysteries in the early part of the 20th century? Apparently foundations and other funding sources have squelched all that—requiring applicants to follow their all-too-restrictive guidelines. As Mark Twain so aptly put it, "What gets us into trouble is not what we don't know. It's what we know for sure that just ain't so."

In researcher Rex Gilroy's fascinating book *Mysterious Australia* he cites from an old diary written in the late 1800s that a family member inherited, which is now documented history of that region:

> Back in 1898, a Mr. Jack Petheridge was one of a party of graziers in search of good pasture lands beyond Broome in the "top end" of Western Australia on the fringe of the wild north-west Kimberley region. Penetrating inland across the Fitzroy River, they entered the Oscar Range country. Jack was 25 years old at the time and a good shot with a rifle, supplying the group with kangaroo meat during the expedition. What follows is from Jack's own diary, still preserved by descendants now living in Perth.
>
> "My companions and I had been out from Broome for two months, and as we were low on food again I went out one day to shoot more game. I approached a stand of trees and dense shrubbery. When it was but 30 yards distant, I heard rustling among the foliage.
>
> "Then, to my horror, an enormous ape of the gorilla family emerged into view, fully 14 feet in height. His snarling mouth displayed large teeth and his eyes were deeply set within thick eyebrows. His forehead sloped back, and long thick reddish-brown hair trailed from his head, which was sunk into the shoulders, giving him a stooped gait.
>
> "I observed his large genitals and his strong muscular body and arms, which appeared much longer than a normal man's. His hands and fingers were very large and he gripped a high tree-branch with his left hand as he stood looking menacingly at me.
>
> "The man-ape began advancing toward me and it was then that I fired a shot at the brute's chest. He screamed and clutched his chest, but kept coming so I fired again—a fatal shot at his head—and brought him down only feet from me. The man-ape was covered over much of his body in thick reddish-brown hair and had very large feet with an opposable big toe.
>
> "I ran back to camp to tell my disbelieving companions, but after they saw the body, the first thought was how many more of these gorillas were thereabouts. But the creature's great height and bulk was much more than any ordinary gorilla to

> our knowledge and, anyway, what were such animals doing in Australia?"
>
> The men left this "gorilla" lying there and abandoned the region to head for home. Jack later returned to the area with a naturalist, but by then, months had passed and they could find no trace of the animal's bones.
>
> There are many historical accounts preserved across Australia, all of which demonstrate that our pioneers took the existence of the "hairy men" very seriously.[2]

Nowadays, the giants retreat from us and hide because there are more gun-happy researchers, deer hunters, etc. But again, the killer-advocates are so sure they are after a wild untamable beast that they put Sasquatch on the defensive during the search. My peaceful approach to field research has proven most successful.

In 1979 I developed a unique and viable field methodology that successfully led to major contact with these sentient nature people. I am the first scientist to discover the following:

1) The Sasquatch are authentically a "people" and not animals.
2) The Sasquatch have a genuine and profound psychic ability that they can use as a survival mechanism against aggressive modern man by dematerializing.
3) These advanced humanoid beings, though they are hair-covered, have language and culture using mental telepathy to communicate with their own race as well as a select few people in the outer world in rural areas. They can speak verbally as well.
4) There are four separate races of them on the North American continent: the ape-like ones, the dog-faced ones, the Ancient Ones with a human face, and the orangutan-like ones in Florida. All are humanoid.
5) They are affiliated in some way with friendly ETs/UFOs.
6) Some of the Ancient Ones have the ability to read and print a poor form of English and artistically use traditional First Nation Native American signs and characters. This was discovered and documented in 2004.
7) As of January 1, 2011, I have personally documented 187 paranormal encounters that report mental telepathy, interdimensionalism, and some ET/UFO activity.

8) In the last 32 years I have had seven physical sightings, plus well over 1,000 telepathic contacts with the Sasquatch people and Ancient Ones.
9) These beings told me in 1985 that they were brought here and "seeded" long before the human race (*Homo sapiens sapiens*) was also "seeded" on this planet after being genetically engineered.
10) As of 2009, I began teaching others my benign, nonviolent approach to making contact, and have shared the method with several contactees.

The 2003 book *The Synchronized Universe: New Science of the Paranormal* by Massachusetts Institute of Technology and Princeton University physicist Dr. Claude Swanson reveals that the tapestry of modern science is showing a few tatters. Swanson lists "12 Things that Science Can't Explain…But That Happen Anyway." These unfolding mysteries point the way to a new, deeper science that no longer denies spirit and consciousness, but acknowledges and embraces them. The difference with the Sasquatch and ET/UFO enigmas is that, because of our lack of civilized behavior on this planet, these evolved races of beings can manipulate us and the circumstances surrounding us. The Sasquatch and UFOs should have been 13 and 14 on his list, but there is enough data here to indicate that many mysteries are literally at the threshold of being solved. In the past three decades scientific evidence has accumulated showing that the present scientific paradigm is broken. In the hard sciences:

1) *Dark matter* of an unknown form makes up most of the matter of the universe. This matter is not predicted by the standard physics models. The so-called "Theory of Everything" does not predict and does not understand what this substance is.
2) *The law of gravity* appears to be seriously broken. Experiments by Saxl and Allais found that Foucault pendulums veer off in strange directions during solar eclipses. Interplanetary NASA satellites are showing persistent errors in trajectory. Neither of these is explained or predicted by the standard theory of gravity know as Einstein's General Relativity.
3) The *cold fusion* phenomenon violates physics as we understand it, and yet it has been duplicated in various forms in over 500 laboratories around the world. Recent

studies by the Electric Power Research Institute, a large nonprofit research organization funded by the nation's power companies, found that cold fusion works. A recent Navy study also verified the reality of cold fusion, and the original MIT study that supposedly disproved cold fusion has been found to have doctored its data. Present day physics has no explanation for how it works, but it does work.

4) *Charge clusters*: Under certain conditions, billions of electrons can "stick together" in close proximity, despite the law of electromagnetism that like charges repel. Charge clusters are small—one millionth of a meter in diameter—and are composed of tens or hundreds of billions of electrons. They should fly apart at enormous speed, but they do not. This indicates that our laws of electromagnetism are missing something important.
5) *Cosmology*: Quasars, which are supposed to be the most distant astronomical objects in the sky, are often found to be connected to nearby galaxies by jets of gas. This suggests that they may not be as far away as previously thought, and that their red shifts are due to some other, more unusual physics that is not yet fully understood.
6) *Speed of light*, once thought unbreakable, has been exceeded in several recent experiments. Our notion of what is possible in terms of propagation speed has been changing as a result. Certain phenomena, such as solar disturbances on the Sun that take more than eight minutes to be visible on the Earth, are registered instantaneously on the acupuncture points of instrumented subjects. Acupuncture points apparently respond to solar events by some other force which travels through space at a much higher speed than light.

This covers just a few of the more glaring anomalies in the "hard sciences." Evidence has also accumulated in the laboratory that many paranormal effects are real, and can be verified and studied scientifically. Among these are the following:

7) *Extra-sensory perception:* Large-scale experiments by the Princeton PEAR Lab and other laboratories have proven that ESP is a real, statistically verifiable scientific phenomenon. Thousands of experiments have been conducted with dozens of subjects, which demonstrate that this form of

communication is real, and that it does not weaken measurably with distance. This makes it unlike any known physical force.

8) *Psychokinesis, or mind over matter:* The ability to exert psychic force over objects at a distance has also been demonstrated in large-scale experiments. Even over distances of thousands of miles, the behavior of random event generators has been altered by the intention or the psychic force of a distant person. The odds that these effects are real, and not due to chance, are now measured in billions to one. In other words, this phenomenon is real.
9) *Remote viewing*: The American military conducted a secret remote viewing program for almost two decades. It was supported because it worked, and evidence of its success has now become public. The remote viewers have demonstrated that it is possible to view "targets" which are remote in space and time. In many cases details that were unavailable any other way were acquired by the viewers. Rigorous statistical experiments have confirmed that remote viewing has accuracy far above chance, and represents a real phenomenon that defies present scientific explanation.
10) *Time and prophecy*: One unusual aspect of ESP, remote viewing, and psychokinesis is that "time" doesn't seem to matter. One can exert an influence or acquire information in the past and in the future almost as easily as in the present. In conventional physics, the order of events is very important, but in the realm of psychic phenomena there seems to be a flexibility to move in time that defies current physics.
11) *Out-of-body experience*: Experiments have been performed which show that during some out-of-body experiences, the "astral body" or center of consciousness of the individual can be detected at remote locations. When individuals go "out of body" and focus their consciousness at another location, physical disturbances have been measured at that remote location. These include anomalous light, electrical, magnetic, and other physical forces, which indicate that the "astral body" sometimes has physically measurable properties.
12) *Ghosts*: Modern scientific ghost hunters use magnetic,

> electrical, optical, and thermal sensors when they survey supposedly haunted sites. In hundreds of cases, technically trained researchers have found measurable physical anomalies when ghosts are said to be present. Although some people have claimed to see ghosts, and many have reported anomalous cold spots and described a strange chill on their skin, modern ghost hunters have shown that unusual magnetic fields and strong voltages also occur in these same haunted locations. Unusual orbs have been photographed at the same time that magnetic and electrical disturbances are measured. None of these can be explained by conventional science.[3]

The above list gives one a clearer idea of just where science is now, and indicate that an eclectic, interdisciplinary approach to the Sasquatch problem is more productive than a rigid zoological one. Sasquatch researchers as well as nature-lovers can begin to increase the amount of contact, connecting with the Sasquatch people by merely following the spiritual protocol in this book. For some, this may be difficult; for others it will be easy. Yet, all difficult tasks can be rewarding, if one sincerely and with great patience works at it.

Clearly, with the super-uniqueness and complexity of this aspect of cryptoanthropology, it is imperative that we move beyond Darwinian biology, beyond Newtonian physics and Freudian psychology, into the nether-dimensions of quantum mechanics, humanistic psychology, and even into the spiritual realm of the ancient avatar. This is where truly advanced thinking of science lies in our future.

In Ervin Laszlo's brilliant and urgent book *Quantum Shift in the Global Brain: How the New Scientific Reality Can Change Us and Our World* (2008), it is stated:

> Science's cutting edge now views reality as broader, as multiple universes arising in a possibly infinite meta-universe, as well as deeper, extending into dimensions at the sub-atomic level. Laszlo shows that aspects of human experiences that had previously been consigned to the domain of intuition and speculation are now being explored with scientific rigor and urgency. There has been a shift in the materialistic scientific view of reality toward the multidimensional world view of multiple interconnected realities long known by the world's great spiritual traditions.[4]

In a far less wordy and more sophisticated way the Sasquatch people speak of this important *interconnectedness* that Old World science and the public at large has ignored or been unaware of. This interconnectedness eradicates much of the ego, allowing us as an evolving modern society to see the Sasquatch, Ancient Ones, ETs, and other races and species as integral "neighbors" to learn from—not to loathe, fear, and exterminate, for to destroy them would be to destroy a part of ourselves. The Native Amerindian philosophy about us being within the Oneness of Nature is, *for real*, a proven aspect of modern physics as interpolated by author Ervin Laszlo within the field of quantum mechanics. *This is an extremely critical concept for mainstream science and bigfoot researchers to understand.*

In essence, this concept could create "the hundredth monkey effect." Hypothetically, if field research were carried out based on sharing, trust, kindness, camaraderie, and love toward the Sasquatch people (and ETs), they would reveal themselves without fear of being tricked, captured, molested, dissected, and exploited. They would be their own "proof" in the flesh. This would create a whole new precedent. The forest giants would share their wisdom rather than be set up to be studied. The Sasquatch people have been studying *us* and our dysfunctional culture for years; we have been monitored all this time and not harmed by them in any way. Yet our approach in seeking them out has been based purely on a dogma of proof, violence, and profit. These are not ethical principles of a modern society, so why would the giants acquiesce?

Going back to the all-important message of Laszlo, because it directly applies to every human and humanoid being on the planet.

> Scientists now know that particles are entangled—non-locally connected—with each other throughout space: they have a prior unity that is active and manifest. Living things of all kinds are non-locally connected throughout the biosphere; theirs is a subtle connection that is likewise active and real, although we have only recently discovered it. As anthropologists have found, so-called primitive but in many respects highly sophisticated tribes are also non-locally—telepathically—connected with one another, with their homeland, and with their environment. They have not repressed their prior unity. But modern humans have repressed the recognition of our prior unity and then, emboldened by misguided rationality, denied its very existence.[5]

Just who is this person who speaks about scientific proof and the profundity of interconnectedness in our presently fragile planet, the need for humans to wake up to a greater spiritual awareness as a part of a cosmic plenum, and the "new" fundamental concept of reality if we are to survive? Ervin Laszlo is the editor of the international periodical *World Futures: The Journal of General Evolution*. He is also Chancellor-Designate of the newly formed Global Shift University. Laszlo, twice nominated for the Nobel Peace Prize, is also the author of 83 books, translated into 21 languages. He is the founder and president of the Club of Budapest (*www.clubofbudapest.org*), which has an extremely long membership list of distinguished spiritual leaders and intellectuals that includes the Dalai Lama, Dr. Jane Goodall, Russian statesman Mikhail Gorbachev, scientist/astronaut Dr. Edgar Mitchell, and many other dignitaries from all over the globe.

Ervin Laszlo is proposing a new world morality, based on the newly discovered aspects of science, which include the very real physics that explains interconnectedness. Since this interconnectedness is already a scientifically proven fact and not just an ancient philosophy, then harming a Sasquatch in any way is actually harming ourselves. With every action there is an equal "rebalancing" reaction! The Sasquatch are evolved beyond these treatises, and they know that *Homo sapiens sapiens* is on an eco-socio-political suicide trip of global proportions. The elders of Sasquatch tribes are probably the wisest beings on the planet and understand the clandestine schemes from the White House to the Russian Kremlin.

Laszlo says about our possible evolution:

> Human consciousness is not a permanent fixture: cultural anthropology testifies that it developed gradually in the course of millennia. In the thirty- or fifty-thousand year history of modern human beings, the human body did not change significantly, but human consciousness did. How will it change next? The answer to this question is of more than theoretical interest: it could decide the survival of our species.[6]

Be the change you wish to see in the world.
Mahatma Gandhi

CHAPTER 7

A NATIVE AMERINDIAN PERSPECTIVE

In 1979 when my first contacts with space beings and Sasquatch began, I was working in applied anthropology with the Great Lake Woodland Indians in an administrative position in community health and human service planning. During that time, I met two traditional medicine people who shared their knowledge of Mother Earth, herbal medicine, shamanic ceremonies, the Sasquatch, Starpeople, and Native American religio-philosophy. I soon discovered that many of the things labeled folklore and legends in anthropology textbooks were, in fact, genuinely authentic events and situations of a nonphysical reality.

Before experiencing the nature of the psychic Sasquatch, I too was a skeptic who rigidly adhered to the principles (and limitations) of empirical science. So it was not easy for me at the time to accept the new but ongoing visits from other-dimensional beings. It was the quiet sharing by my trusted Indian friends of their world of sacred interdimensional thinking that, over time, dramatically changed my life forever.

In the spring of 1981 I was invited to study the Old Ways with an Ojibway shaman named Keewaydinoquay (Woman of the Northwest Wind) on a remote wilderness island in the western Great Lakes. Grandmother Kee was very traditional in everything she did, but she also had a PhD in ethno-botany. She had a foot in two worlds. "I have seen the Starpeople walk by me several times in the forest while picking herbs as well as seen their ships hovering over the trees. They use a 'doorway' on the island that leads to another world," Kee related. Once she said she ran when startled by a twelve-foot-tall "Bugwayjinini" (Bigfoot). Often she would say a word in the Ojibway language for an herb or other item, then immediately tell me in English. It was a place to acquire more survival skills, understand Indian herbology, and a time for esoteric learning.

I felt very fortunate to have the opportunity to live on such an isolated island for five months. It intrigued me. I thought back to a 1969 adventure when I boarded a French freighter in Australia and spent nearly two months island-hopping across the South Pacific. But on this particular island, the venture was very different. We spent hours talking, studying plant identification, and gathering wild foods on our frequent field trips throughout the island woodlands. Kee expounded on how spirits could aid in natural healing. Some of what she shared was akin to the series of books by Carlos Castaneda.

My mentor believed that Bugwayjinini was a spirit that temporarily entered the material plane and, since it was a spirit, could not be killed. This was her way of understanding Sasquatch's elusive nature. Once, during a ceremony she called in a group of Indian spirits. I assisted her as asked, going along with what I thought was pure folklore. But when Kee commanded that the spirits come "*now*," I was astonished to actually feel invisible entities pushing past me on both sides. There was no mistake about what I had experienced. From that time on, I stopped labeling these Native American rituals folklore.

After living with Keewaydinoquay for one month, I had an Indian fisherman take me to a more isolated island to continue my botanical studies alone. Kee said it had resident Sasquatch and occasional Starpeople there. The Indians spent very little time on this island because of its reputation for being mysterious. They spoke of it with apprehension. I was to camp there for four months without any contact with the outside world.

I located an adequate campsite on the island, in a field with a large old-growth hardwood forest to the rear. Three hefty army duffel bags of food were hung from trees to protect the contents from rodents and other animals. Water was hauled from the lake a quarter of a mile away and purified before drinking. Wild food was abundant and easily supplemented my rice, dry soups, dried fruits, granola, peanut butter, whole-wheat crackers, and canned fish. Every day I gathered a variety of wild edibles made up of roots, stems, leaves, and flowers, and made a huge salad to eat. An array of wild berries and mushrooms was accessible for nearly two months. In the autumn I picked wild apples.

Overall, I did not get bored with my diet because of the diversity of wild foods available to me. Sometimes I soaked red staghorn sumac buds for an hour or two, then strained the water through cheesecloth to catch the fuzzy fibers. This concoction was a kind of Indian "pink lemonade" that was delicious and nutritious.

Occasionally I made an improvisational cheese by boiling powdered milk, adding finely chopped wild leeks for taste, then dropping a handful of cleavers (a wild herb) into the heated pot. In a short while a portion of the skim milk would coagulate, forming curds and whey. The solution was then strained through cheesecloth and the curd shaped into a ball, which I wrapped in the cloth. It was finally hung on a tree branch to solidify into cheese. Crackers and sliced cheese would be on the menu the next day. Such food gathering and preparation acted as a peaceful pastime that was deliberately meant to attract Sasquatch curiosity. During my daily walks I casually looked for spoor, observed an abundance of wildlife, and identified edible plants that might possibly be harvested by a Bigfoot creature. There were ample amounts of food available for any herbivorous/frugivorous biped.

Each day, I sat for hours in an eight by ten walk-in tent, reading journals, magazines, books on ethology, botany, anthropology, ecology, zoology, native American studies, biographies, history, Eastern philosophy, and parapsychology. I also organized and wrote copious field notes. Plus, I felt that a vital factor in my field research was one hour of meditation twice daily, followed by mental telepathy. With this technique, I projected loving thoughts to any Sasquatch in the area. This helped me to flow better with the Oneness of Nature in hopes of building trust. Such an approach was purely experimental at the time.

Wallace Piawasit, a traditional Potawatomi medicine man and elder, advised me before I left Milwaukee, Wisconsin, to thank the Great Spirit every evening at sundown for all the wonderful experiences of the day. He told me to face west, speak aloud, and sincerely express my gratitude to all the trees, flowers, animals and Our Earth Mother. After the first few days of doing this, I began to feel serene and more connected to nature. A pair of rabbits appeared and frolicked at my feet, sometimes touching me. Two woodcocks would land in front of me on the ground, always about three feet away. Once I started praying aloud, these animals would inevitably join me every evening for twenty minutes until I finished with prayer. It amazed me. One evening a whippoorwill landed on my head in front of my little audience of regulars. Another evening an owl attempted to land on me while ignoring its usual prey that was huddled around me! This was a powerful lesson for me. It allowed me to get outside of myself, let go of ego, and just flow with unconditional love—very simple but profound.

One whole month passed before anything unusual occurred. Not a single sign of Sasquatch anywhere. Then something bizarre and puzzling began to happen. I still do not completely understand what transpired, but will relate what happened in the event a reader has had a similar perplexing experience.

One day at camp I was stunned—completely unnerved—by the enormous crash of a large tree falling to the ground! It appeared to be about a hundred feet behind my tent on a quiet, sunny, windless afternoon. I had an 8mm zoom-lens movie camera at the time, which was in hand as I tiptoed through the brush to the area where I had heard the horrendous crash. Nothing! No trees were downed. This deeply puzzled me. Then I heard the sound of "someone" with a long stride slowly walking away. It sounded like it was 30 to 40 feet in front of me, but nothing was visibly there! Since that time I have experienced this phenomenon often, as one of the mysterious behaviorisms that characterize the psychic Sasquatch.

Over a five-week period I counted fifteen times that I experienced the "gigantic tree crashing phenomenon." Sometimes it was about two hundred feet from the tent; other times as close as fifty feet. Each time it was as if a 50- to 60-foot-tall tree had fallen, making a tremendous noise as it struck the ground. Yet, whenever I investigated there was *never* any sign of a fallen tree! At times, seconds after the crash, I would hear a bipedal "person" walking away. During the four months on the island I never once had a visual sighting of a Bugwayjinini. However, I did have a daytime sighting of a spaceship moving slowly over the central part of the island. The "disk" was flying low, perhaps 400 to 500 feet up. With my zoom-lens binoculars, I had an excellent view of the craft. The sighting lasted approximately one minute. I was so excited to see the object that I never ran back to my tent to get my movie camera.

I experienced a similar the tree-falling episode in Texas in 2004, while camping in the woods. I was having almost daily visitations from the Ancient Ones. I was leaving food and crystals for them and they would leave a rock for me, plus one of them would usually print out a note for me on a pad beside the food. It was fun.

One night around 10:30, while I was curled up in my sleeping bag and almost asleep, there was a terrific "crash" from a large tree only 20–25 feet from my tent! I nearly jumped out of my sleeping bag with fright, since it caught me completely by surprise. Then I heard the familiar bipedal steps walking away. The next morning I immediately surveyed the forest, meandering out to 50–60 yards in diameter. Not one downed tree could be found. The "tree" seemed like it had fallen beside the tent by the sound of it. So I must surmise that these spectacular tree-crashing sounds were *somehow* psychically produced by the forest giants. The Texas incident was the sixteenth time I experienced it and I have no other conclusion at the time of this writing. There is no empirical evidence to explain these anomalies.

The psychically gifted Sasquatch have strange abilities beyond what strictly rational humans can comprehend. The giants are highly advanced humanoids living inside hairy animal-like bodies. Their status is confusing to us because we have been led to believe there are either human beings or animals on Earth, with a rigid line drawn between the two. To understand and accept these forest people, we must think outside the box by realizing, as they have told me and others, that the man-creatures evolved on another planet, migrating here eons ago, via their friends the high-tech Starpeople. Like me, many Amerindian people know the truth of the psychic Sasquatch and respect them as wise beings with immense powers.

Native Amerindian people would be in agreement with biologist Lyall Watson when he stated, in his book *Supernature* (1973):

> There is life on Earth—one life, which embraces every animal and plant. Time has divided it up into several million parts, but each is an integral part of the whole. A rose is a rose, but it is also a robin and a rabbit. We are all one flesh, drawn from the same crucible—this is the secret of life. It means that there is a continuous communication not only between living things and their environment, but among all things living in that environment. An intricate web of interaction connects all life into one vast, self-maintaining system. Each part is related to every other part and we are all part of the whole.[1]

First Nation people have no need to capture and dissect their "big brother." When will we as a society learn to respect the living world around us from a spiritual standpoint and not just an empirical one? The Sasquatch people are waiting for us to learn this, but time is running out, as we are preoccupied with raping our precious environment and always preparing for war.

In August 1976, while investigating Sasquatch encounters in western Montana, I met a Cree Indian from Canada. He told me he knew a medicine man in Alberta who often met with a Sasquatch tribe high in the Rockies. The giants would take the shaman to a cave where they would communicate "mind to mind," and trade herbs. The Indian lived in the valley and the hairy folks in the high country. The medicine man would pick herbs that did not grow up high and so they had a fair exchange. The Cree Indian said that one day while in the wilderness with his medicine man friend, he saw a group of Sasquatch high above them on a ridge, pulling a dead elk by the horns. Later, the medicine man told him that the Sasquatch are a "people," not animals as the White man believes.

In January 1981 I met a First Nation medical doctor outside of Spokane, Washington. He was retired and had returned to his traditional ways, merging the ancient values with modern techniques. The Indian doctor told me that while he was delivering a traditional prayer ceremony to a group of First Nation people in Canada at night next to a large bonfire, a Sasquatch had telepathed to him saying it wanted to join them in prayer. In his mind, the doctor told the Sasquatch he was welcome to participate. Then the doctor interrupted the ceremony to tell the group that another "person" would be joining them and they were not to be alarmed when he arrived. A short while later, the man-creature stepped out of the darkness of the forest, standing only a few feet behind the Indians in the prayer circle. After the ceremony was completed, the giant quietly walked back into the night. This happened on two separate occasions. For many First Nation people a Sasquatch visit is considered a sacred event.

While working at an urban Indian agency in Wisconsin, I met a medicine man from Wyoming who told me that because there is less forest and more rolling hills in that state, the Sasquatch often travel in the deep gullies to avoid being seen. The man told me that his sixteen-year-old grandson had been out hunting with a .22 caliber rifle and had followed a gully looking for deer. He had squatted down to examine some tracks and, when he stood up, he was startled to see a seven-foot Sasquatch standing a few feet away, starring intensely at him. In fright, the young man dropped his gun and tried to climb up a steep embankment, but kept slipping back down. The creature stood nonchalantly observing the terrified lad.

Later, the boy was found walking in a daze, unable to speak. He was in a state of shock to the point of being catatonic. Finally, he was taken to a larger hospital in another state for observation. Five days later, the boy came out of the stupor and eventually told his story of meeting the Sasquatch. The Indian boy said the creature did something inside his head that mentally incapacitated him, putting him in a trance. The White doctors did not believe him, but the young man's Indian family believed that what he related was true, because they knew the giants had unusual powers.

In the article "On the Cultural Track of the Sasquatch" found in the book *The Scientist Looks at the Sasquatch* (1979), anthropologist Wayne Suttles wrote from his field notes a statement by a Lummi Indian elder when he said the Sasquatch

> ...is a great tall animal or whatever it was that lived in the mountains. It was like a man but shaggy like a bear, like a big monkey 7 feet tall. They went away when the Whites came. The Indians never killed any; it was pretty wise animal, or whatever you call it. If you saw one it made you kind of crazy. They throw their power at you.[2]

A few people have personally told me that the Sasquatch tried to control their minds during an encounter. This also happened to me on four separate occasions. I had a weird feeling in my head and could not move any part of my body during the mental invasion. But they never did it again after I asked them to stop in the mid-1980s. I broke the "spell" in an instant by saying, "May the White Light of Christ surround me and protect me." I never found out what their purpose was in doing so, but they respected my wishes the very first time I requested that they stop. I have a completely different relationship with them now, and my life is richer for it.

The forest beings have on every occasion been kind and responsive to me—a few times referring to me as their "brother." Once a Sasquatch referred to me as a "doctor" because I am a trained herbalist. The Ancient Ones affectionately call me "Dream Man." The Bigfoot people I have encountered have values that parallel the Amerindian cultural practices, because both races live in nature, hunt and fish with respect for the land, and are very family oriented. I have found that, if you *sincerely* reach out to them as a friend, they will often help during times of inner strife. Communication is the key. They know a person's intentions and will choose to appear or interact when they feel a person is ready. It is always they who control this.

I have noticed that there are more Sasquatch people on Indian reservations, because it is a sanctuary away from the sacrilegious White man. Native people living in the traditional way do not normally shoot at or molest the forest giants and white Westerners are not allowed to hunt or trespass on Indian land.

Tribes in different geographic areas have different "cultural" beliefs and values regarding the Sasquatch, and each has a name for them in the native language. All tribes say the beings have special powers beyond what mortal man understands.

Some Indian nations saw the Sasquatch as a good omen when it visited their area. For example, fire spotter Willard J. Vogel (now deceased), who worked on the Yakima Indian Reservation in Washington State, wrote about the Yakima Indian "legend" of a large man with red eyes who came to live with the tribe long, long ago. The giant did many things for the tribe, including healing the sick. One day, the Sasquatch knew he was dying and asked the Indians to take him to a certain location in the forest. Soon after being taken to that spot, he died. Then a large flying object landed, the occupants took his body aboard and flew away.

Many Indian tribes view the Sasquatch as a "big brother" (and Sasquatch have referred to me as their "little brother" at times).

The giants come to give counsel, spiritual messages, or warnings. Author Preston Dennett in his 2006 book *Supernatural California* says that a few Indians in that state slowly began to share some of their encounters with the local White man.

> In the summer of 1897, a Native American who lives in Tulelake reported that he was approached by a Bigfoot while fishing. He offered the Bigfoot one of the fish, which it accepted. The Native Americans report other friendly encounters. In the early 1900s, a native of Mount Shasta reports that he was hiking in the wilderness when he was bitten by a poisonous snake. Incredibly, he was then approached by a Bigfoot, which lifted him up and carried him back to his camp.[3]

However, some tribes have had great fear of the Sasquatch. Indians have told of being cannibalized by the giants in the distant past, and it has been passed down through Indian oral tradition that cannibalism was wide-spread among the different Sasquatch tribes. Still others view the hairy folks purely as tricksters who steal food and kidnap women and occasionally their children, which the Sasquatch make part of their families. Some stories claim that the victims are used as slaves.

Many of the Indian tribes spoke of the giants as "demons" because of their mischief. In the well-documented book *Giants, Cannibals, and Monsters: Bigfoot in Native Culture*, anthropologist Kathy Moskowitz Strain cites from "Indian Legends of the Pacific Northwest" of the Puyallup Indians:

> The *Seatco* were neither men nor animals. They could imitate the call of any bird, the sound of the wind in the trees, the cries of wild beasts. They could make these sounds seem near or seem to be far away. So they were often able to trick the Indians.[4]

Numerous tribes tell of the Sasquatch's ability to become invisible, especially when it's hunting, as this confuses the game. Personally, I have seen twice a Sasquatch blip into another dimension. One minute it is physical, the next—invisible. Many times I have had stones and sticks gently tossed in front of me, but when I looked in the direction from which it came (ten to fifteen feet away), there was nothing visible. It was their playful way of letting me know they were there.

Another Native Amerindian perspective of reality regarding

Sasquatch-type beings is their ability to "shape-shift." In Michigan and Wisconsin, and no doubt in other states, witnesses have reported a dog or wolf shape-shifting into a man or wolf-man—sometimes referred to as a "werewolf." Somehow, these two states have more reports than any other concerning this fantastic phenomenon.

Conservative Bigfoot researchers scoff such an idea. But in the 1970s I scoffed at the Bigfoot/UFO connection, interdimensionalism, mental telepathy, their ability to walk through solid mass, psychic healing powers, their ability to "aport" objects into this dimension, and the view that the Sasquatch are an evolved nature-people. I discovered all of the above are absolutely *true*! I have personally witnessed these psychic characteristics in the Sasquatch more than a thousand times over the years.

Amerindian legends and traditions tell of the sacred white buffalo and the spiritual meaning behind it. The Plains Indians say that a rare white buffalo transformed into a beautiful woman, offering to the indigenous people the gift of the sacred pipe ceremony. Interspecies communicator Kathleen Jones had a herd of white buffalo on her Oregon ranch.

The sacred White Buffalo at Kathleen Jones' ranch in southwest Oregon

The Sasquatch people told her that they honor the animals and would protect them.

Haloti, the Ancient One in Oklahoma, told me that a few people in her tribe have the ability to shape-shift! I have no reason to doubt her, as she has been consistent with other statements to me. They and the Sasquatch people have tremendous mastery over the physical world. It borders on science fiction. That's why friendly communication with them is key to understanding the Sasquatch mystery. The control of the psychic realm by the forest giants is actually a scientific discovery I made that can be proven when researchers begin using psi in the field, starting with basic mental telepathy. The more people who experience the psychic Sasquatch, the more their abilities will become an accepted fact.

Many First Nation medicine people have the ability to shape-shift, but unlike White people, Indians have no need to come forth to be put on exhibition to prove anything. An acquaintance told me about a White woman who was driving through the Navajo Reservation and stopped to pick up a handsome and nicely built young Indian man hitchhiking on the road. The front passenger seat was cluttered, so she motioned for him to sit in the back. As they began to make conversation, she would periodically glance in the rear view mirror at him. After a time, the young man indicated where to drop him off. As the woman approached the drop-off point, she fell into a state of shock to see through the mirror a very old Indian man smiling back. He laughed while thanking her for the ride. Apparently the elder used his powers to transform himself into an attractive person that would assure him a quick ride. To my knowledge, this is a true story. If some humans have the ability to shape-shift, then certainly some of the hairy folks do too.

A comment posted by a person named Chaska Denny on the website *www.bigfootsightings.org* on June 21, 2007, is worth sharing:

> I am Native American, raised up by my Grandfather and Grandmother, not by circumstance, but by choice, they took me everywhere and I was witness to the "Old Ways" of our Native People on our Mother Earth. Shape Shifting is a Traveling Form used by those who know the Medicine Ways—Medicine Lodge societies. I have witnessed this and know how powerful it is, yet it is something normal to the Medicine People, it is their way of traveling long distances to gather certain herbs, roots, and natural medicines. They can assume animal shapes for shorter

> distances and also become a pure sphere of light for longer distances, about as big as a common basketball, pulsating like breathing. The ones called Bigfoot have many powers besides shape shifting, they can read a person's mind, they can communicate mentally with each other and to certain people who are sensitive. They know of this world and what is going on with us, they avoid what we do and keep themselves hidden, unless they want you to see them, then they show themselves. Is it no wonder *why* none have been captured? Because they know your thoughts and can read thoughts in about a 5-mile radius.[5]

Mainstream America is lagging far behind when it comes to spiritual/psychic ways. Many persons relegate such things to science fiction or Old World superstitions instead of learning from real ancient wisdom. Establishment scientists reinforce this negative view. The Sasquatch have said humans that are too self-centered, too judgmental, too wrapped in technology, that we don't listen, and are therefore evolving more slowly.

On July 16, 2007, Chaska Denny shared more about the psychic Sasquatch:

> …as I stated, the medicine way of shape shifting has been around for eons. I do know that shape shifting isn't only localized to a particular area or race, it is not something spoken about by other cultures, except in certain stories or legends. If one looks at these stories/legends, then it is easy to see that Medicine people of that culture or race were probably involved. I have witnessed shape shifting and it's something that shook my way of thinking and understanding completely, as my elder said: "We are going to wash your brains"…they did…so I have learned not to limit myself and thinking to what I thought I knew or what may be written. I keep an open mind and look and try to understand it in a new way. We call "Bigfoot" our "Elder Brother," we know Creator created them for a purpose, they come and go as they please, so just as we cannot see the *air* we breathe, that doesn't mean it doesn't exist.
>
> In the 70s there were reports and sightings of *bigfoot* in the Little Eagle, South Dakota area. Our Medicine men were contacted and requested to look into the matter. They held a ceremony for 4 days in the area. In the dark of the moon, all of

> a sudden *he* came, without a sound, and communicated to them with certain sounds, but mostly with mental telepathy. "They" have always been here and are spread out all over the world. They watch over forest areas; they have helped us many times in the past. They are moving here and there now, because their habitat is being invaded by those who cut the trees and dig into the mountains. "They" give warning to those who try to find and track them. There is nothing to fear *if* a person does not try to interrupt them or prevent them from moving about. They stay hidden because it is their nature and do not want to be seen by people unless they decide to let themselves be seen and there is a purpose in letting themselves be seen.[6]

I respect the Native Amerindian perspective as they themselves historically were a nature people who knew all the animals, trees, waters, and forest beings in their environment. Who are the Whites—as outsiders—to criticize their ancient knowledge? As Native American attorney and professor, Vine Deloria, Jr., relates in his 1997 book *Red Earth, White Lies*:

> Regardless of what Indians have said concerning their origins, their migrations, their experiences with birds, animals, lands, waters, mountains, and other people, the scientists have maintained a stranglehold on the definitions of what respectable and reliable human experiences are. The Indian explanation is always cast aside as a superstition, precluding Indians from having an acceptable status as human beings, and reducing them in the eyes of educated people to a pre-human level of ignorance. Indians must simply take whatever status they have been granted by scientists at the point at which they have become acceptable to science. There was a terrible reaction in 1969 when I accused anthropologists of treating Indians like scientific specimens.[7]

Not only is this vital educational cultural information a great insight about First Nation peoples, but when we include the Sasquatch peoples in the overall scenario, it becomes clear that the dominant Western culture in North America is still perpetuating racial prejudices. Indians view the giants as a people, but White society does not. It appears that much of mainstream America (and Canada) makes the same cultural mistakes over again, while thinking of themselves as superior and "civilized"! Deloria continues his point by saying:

> The stereotypical image of American Indians as childlike, superstitious creatures still remains in the popular American mind—a subhuman species that really has not feelings, values, or inherent worth. This attitude permeates American society because Americans have been taught that "scientists" are always right, that they have no personal biases, and that they do not lie, three fictions that are impossible to defeat.
>
> Scientists may not have intended to portray Indians as animals rather than humans, but their insistence that Indians are outside the mainstream of human experience produces precisely these reactions in the public mind.[8]

The Sasquatch people have repeatedly told *many* different sources, including myself, that White society is a most disrespectful and destructive group of people, whose negative energy toward living nature

The author on a 1980 expedition in the Columbia River Gorge with a 110-pound backpack. He saw Mt. St. Helens erupt from his wilderness campsite.

may cause them to self-destruct. They say that our Mother Earth is in peril, yet we are like a racing locomotive that is blindly coming to the end of its track!

Other interdimensional forest beings also exist in nature and are an integral part of Indian culture. The "little people" are not small ETs; they are entirely different. When I asked Haloti about the little people, she said that every autumn the Ancient Ones have a corn festival during which they pick the last of the corn from a farmer's field. Then the Sasquatch and little people gather to join the Ancient Ones to eat and trade. Haloti said the little people mine gemstones and there is an exchange of items amongst them. It is reminiscent of Snow White and the Seven Dwarfs, in which the dwarfs work in a mine digging up gemstones. So maybe there is an element of truth to these children's stories. Many tribes speak of the little people, who have mostly a positive image. However, some tribes avoid them and view these beings as a bad omen.

While in Oklahoma in 2004, I map-dowsed the entire southeastern part of the state for a friend to see if there were any little people. I located one region, pinpointing the spot where he should to go. He made several visits to the place and had the privilege of encountering the little people twice! All he had with him was a kind and loving heart.

There are many books written about the little people, and apparently, there are several races of them as well. I suspect they also use "portals," but don't know enough about them to comment further. There seem to be several worlds paralleling ours, since many "now you see them, now you don't" creatures were reported from the time of the early settlers. One should not confuse the little people who wear clothes and look like miniature humans (Leprechauns, et al.) with the four-foot-tall hairy Sasquatch-types.

The only scientific proof for little people was found in September 2003 on Flores Island in Southern Indonesia, when a team of anthropologists found the remains of a three-foot-tall humanoid being! The bone structure indicated that it was female who would have weighed about 30 pounds. Radiocarbon dating of charcoal pieces lying next to the skeleton showed that the "hobbit," as it was coined, was about 18,000 years old. These little people fashioned stone tools, were hunters/gatherers, and had the ability to make fire and cook their food. Scientifically, they were given the label *Homo floresiensis*, although another possible name originates from island folklore—Ebu Gogo. They were supposedly like miniature

Sasquatch of sorts with hairy bodies and long arms. The indigenous people of Flores Island say the Ebu Gogo were a mischievous group that was exterminated a few hundred years ago. However, they could still be extant in hiding or on some of the other Lesser Sunda Islands.[9]

Are they one of the seven races that the Ancient Ones told me are distributed around the globe? Are the Menehunes also of the "race" of Ebu Gogo, but with a Hawaiian name? It's critical to keep an open mind when connecting theoretical dots. The Filipinos have their "Dwende," and there are small hairy bipeds being reported from Mozambique as well as West Africa and Australia. Throughout North America today, tribal Indian elders insist that hairy giants, water monsters, and little people still exist; they don't consider them mythical. Perhaps the seven different races of humanoids have wised up worldwide and remain hidden to avoid perishing at the hands of encroaching modern man.

John Boatman, an Ojibway Indian professor at the University of Wisconsin—Milwaukee, wrote a chapter on extraterrestrial visitations in his book *My Elders Taught Me*. With his permission, I am including the chapter in its entirety, because he mentions "original man" (Bugwayjinini) in relation to the Starpeople. His writing offers additional insights into these important phenomena, because the storytelling is based on real events in Ojibway cultural life.

The Star People

> The Elders sometimes talked about immense flying objects seen very high in the sky which resembled enormous birds and made a sound which sounded like, but was not, thunder. On those rare occasions when they were observed from a closer vantage point, it seemed as if these objects had eyes which flashed like lightning. The Western Great Lakes American Indian tradition has quite a few stories about these "great birds" whose home on Earth was thought to be far to the west. The Elders said that long ago, according to their Elders, a party of hunters, upon entering a large forest clearing, saw an enormous shiny object that appeared to have fog rising from it. The hunters could see flashes like lightning coming from the center of the "fog." As much as they wanted to know what the object was, none of them dared approach it, fearful of what would happen if they did. The next day, however, some of the hunters returned to the clearing to try to determine what it was that they had seen. But again fear

overcame them as soon as they saw the object and once more they left without finding out what it was. When they went back the third day—the object was gone! Where it had been, there was now a large scorched area.

What the hunters had seen, the Elders said, was something they called a "Thunderbird" and is in actuality a relatively small flying vehicle of the Starpeople.

In the summer of 1988, while doing historical research in the Upper Peninsula of Michigan, I had the opportunity to investigate a unique circle of stones on the largest island of a group located between the Upper and Lower Peninsulas. This recently discovered circle, referred to by media sources in Michigan as "Michigan's Stonehenge," has a diameter of approximately 397 feet.

The extraordinary circle was discovered about four years ago by a woman who was a former schoolteacher on the island. I spent some time investigating the circle with her, returning to the site several times before returning to Wisconsin. For the past two years a group of scholars representing several academic disciplines has been investigating this same site.

The structure of the stone circle is remarkable. Unusually symmetrical, there are four immense rocks located precisely at each cardinal point. The colossal center rock has an eight inch hole bored into the top. Another large marker-rock denotes the location of the North Star at the time of the sunrise during the summer solstice. Certain other rock formations and/or locations appear to chart various constellations. Hieroglyphic-like markings and pictographs on several rocks seem to denote astronomical symbols and locations. Test boring throughout the circumference of the circle area demonstrates that there is a large *charred* circular area that does not exceed the perimeter of the stone circle.

When I entered the circle area for the first time, and each time thereafter, I found myself thinking that this circle is the same one that was described in 1856 by Henry R. Schoolcraft in the story, "The Star Family, or Celestial Sisters." While Schoolcraft (probably incorrectly) attributed the source of this story to Shawnee legend, the same or similar story was told by many tribal groups. The Elders from whom I learned said the story was based on a true event that took place a very long time ago on a group of islands located northeast of Wisconsin in the Western Great Lakes.

The following story was told by my Elders and is similar to the one Schoolcraft recorded. Both make reference to a circle like the one on the island in Michigan:

One day a young man who lived with his family in a remote area of the forest wandered into unfamiliar territory and found himself at the edge of an unusual clearing. In the middle of the clearing was a large charred area in the shape of a circle. Outside of the circular area was another circle which appeared to be made by footsteps following the same tracks over a long period of time. The young man knew his discovery was special because he could see no path in the clearing leading to or from the circle.

He would return to this place periodically and conceal himself in the brush at the edge of the clearing while he watched and waited, hoping to discover how this strange circle came to be. One day while he was hiding, he noticed what appeared to be a flash—like that of a twinkling star—in the blue of the afternoon sky. The flash soon became a small, descending object. As it came closer, he heard the faint sound of music. It sounded as if the music were coming from the skies. Looking up once more he saw that the object was coming closer and he quickly realized it was larger than he thought it to be. Soon he understood that he was looking at some kind of sky-craft and that this was the source of the ethereal music that filled his ears. The craft landed in the charred center of the well-worn circle. The young man watched in amazement as twelve beautiful, human-like women descended from the craft, entered the area of the circle, and began to dance to the marvelously enchanting music. When the dancers realized there was someone watching them, they ran back into the craft which started slowly to rise and then suddenly disappeared into the sky. The outline of the charred circle had been freshly scorched by the departing craft.

The young man returned home and told his parents what he had witnessed. The story awakened in his mother a memory of something her grandmother had told her when she was a girl. She turned to her son and explained that he had seen some of the Star People who have very special abilities. When the young man told his mother that he felt as if he could easily fall in love with one of these wondrous females, she felt a twinge of alarm seize her body. She cautioned him that a Star Woman would not be content to stay on Earth. In addition she said that the Star People were not affected by the same time and spatial constraints that humans were, consequently, she would not age at the same rate that he did. She explained that he would grow old and quite immobile, while the Star Woman remained relatively young. Finally she asked him if he felt that such differences could result in lasting happiness. The young man listened and thought for a long time about what his mother told him, but that still could not keep him from feeling drawn back to the special clearing he had found.

Over time the young man observed the phenomenon of the first day regularly. Eventually he won the friendship and finally the love of one of the beautiful female beings who danced to the exquisite music in the clearing. He fathered a child by her and later left with her and the child, bound for her father's home on a planet in the system of a distant star. The Elders said that no one is sure what happened to him. However, some say he, and later his descendants, would periodically return to Earth to observe what is happening here.

The stone circle in Michigan also appears to be related to another story that the Elders told and that Schoolcraft also recorded, under the title "The Son of the Evening Star." The story is about a family in a tribal group living in the Western Great Lakes area.

Living in a village near the lake, there was a family with ten beautiful daughters. The youngest daughter was the most beautiful of all. But she, unlike her sisters, was not interested in the many men who came courting. She instead loved the beautiful, special, and secluded natural places near her home. Even after all of her sisters had married, the youngest and most beautiful daughter continued to ignore her suitors, preferring to spend her time in the special places she loved so much.

Her family often criticized her and warned her that some day she would end up being alone with no chance left of ever getting married.

Then one day an old man, who was scarcely able to walk and obviously very poor, came to call on her. Soon he was visiting every day and it became evident that the youngest daughter very much enjoyed spending her time with the strange and enfeebled old man. Her family simply could not understand her bizarre behavior and, needless to say, they were completely shocked when the daughter announced that she had agreed to marry him!

She insisted that this was her choice and soon after, she married him despite the family's efforts to dissuade her. Many in the village laughed at her and taunted her for having made such a terrible choice for a husband, especially when she had been seriously courted by so many handsome, young, and promisingly prosperous men.

Nevertheless the daughter seemed to be very happy, telling those who laughed at her and taunted her, "It was my choice. You will see in the end who has acted the most wisely."

Shortly thereafter the entire family, including the bizarre couple, was invited to a special feast some distance from their village.

As they all walked along the trail toward the place of the feast, many family members could not help pitying the youngest of the sisters for having made such a terrible choice for her life's mate. This was especially apparent to the family when the old husband kept stopping and looking into the skies while muttering something that was, to the rest of the family, unintelligible. When they looked into the sky to see what he kept looking at, all they could see was a faint flicker that they thought was the Evening Star.

Finally the group came to the place of the feast. They saw that the feasting lodge appeared to be made of some kind of strange metal and was unlike any that they had ever seen before, yet, they entered this strange lodge so as not to offend their host, whom they really didn't know, but about whom they had heard wondrous things.

As soon as they entered the strange lodge, it began to shudder and then seemed to lift away from the ground. The family soon realized that the lodge had indeed lifted off the ground and—not only that—it was moving up and away from the Earth!

The youngest daughter's husband then uttered a joyous sound. When the entire family turned to look at him, they were astonished to see that his appearance had almost completely transformed. Rather than the old, crippled man he had been, he now appeared to be one of the most handsome young men any of them had ever seen. He explained to them that he was not from the Earth and that while on the Earth his physiological processes had been affected by the planet's composition and gravity, which in turn had drastically altered his original appearance.

Finally they came to another world, somewhat like the Earth, but far from the Earth in another solar system. There they learned from the husband's people that some other beings who tended to serve evil had left him stranded on Earth causing his predicament. They also were told that, obviously, their kind needed some additional guidance since, by failing to accept their in-law in the form of the old man, they had demonstrated that they were not sensitive enough to perceive the lack of balance in their own lives. His people explained that a group of beings from their world would be sent with the family back to the Earth to try to provide the necessary guidance needed there.

When it came time to meet those who had been chosen to accompany them back to the Earth, the family saw that their companions were very small. The humans were told that these beings were called the Paueeseegug or "Little People."

They were then directed to follow the Paueeseegug into a craft similar to the one that brought them to the far-away planet. Shortly thereafter, the craft began to rise and the family knew they were on their way back to the Earth.

When the craft was over the upper Lake Michigan area, it began to slowly descend. Finally it hovered over a group of islands located between the present day Upper and Lower Peninsulas of Michigan. The craft came to a stop on the highest of the group of islands. It is said that to this day the Paueeseegug still live in that island group and that sometimes they may even be seen dancing and singing on the moonlit beaches—if they are not disturbed by the humans who are observing them

In the Western Great Lakes American Indian tradition there are many other references in the ancient stories and legends to the Star People who visited and interacted with people on Earth.

The Elders told of other beings who looked somewhat like the American Indians except that they had much fairer skin and were definitely *not* human. The Elders said that these "people" had the responsibility of guarding the heavens in the area of Earth. They were called the "Heaven People" and dressed in scarlet tunics with a hood.

The Elders said that one day an old man of about ninety told the people of his village that he was going to "die" the next day. When asked how he knew, he said that one of the "Heaven People" had told him so in a dream. He told the villagers that they were not to bury his body. Instead they were to take it to a certain island in the big lake. There they were to lay his body on the beach and wait. He said that the Heaven People would come and take it away.

The next morning, just as he predicted, the elder was dead. The villagers did as he had requested, although some of them thought it all seemed quite bizarre. As they waited on the beach of the small island that the old man had specified before his death, they suddenly heard what sounded like thunder. Yet there were no clouds in the sky. Then, the sound of a great wind roared above them.

Soon a strange, shiny craft appeared directly over the beach. From the craft four human-like beings appeared, each dressed in scarlet clothing with hoods drawn over their heads. They approached the elder's body where it lay on the beach. One of them took the hand of the dead body and the old man suddenly rose, looked at the villagers, then smiled and ascended into the craft with the four Heaven People. The craft then slowly rose and, with a loud thunderous sound, vanished from sight.

The Star People also figure prominently in the stories about Nokomis who was the grandmother of the first humans. In these stories and legends it is clear that Nokomis is *not* from the Earth! She and her daughter, the mother of the first human, are said to be from the East, beyond the Earth, from a place referred to as the "Morning Star."

The Morning Star nowadays is identified as the planet Venus; however, in the ancient time referred to here, the Morning Star was identified as the planet between Mars and Jupiter, where an asteroid belt is now. According to several ancient traditions, the belt was created as a result of a massive collision between the ancient Morning Star planet, called Tiamat, and an "intruder" astronomical body. This collision took place during a time referred to in many legends as the time of "The War in the Heavens."

The Elders told of how some of the beings from the Morning Star, who begot the first humans, were offended when later the humans did not remember to honor them appropriately. In their anger they emerged from their base in the west with a noise which reverberated across the heavens. Obscured by clouds, they crossed directly over the homes and villages of the forgetful humans and in their fury they shot what looked like bolts of lightning at the Earth below. Finally, the humans learned to appropriately honor their ancestors from another planet. Although this happened in the distant past, humans continued to symbolically honor the memory of the people from the Morning Star by reverently offering kinnikinnick each time that a thunder storm rumbled through. The storm sounded like the crafts of those ancient ancestral beings.

The Elders stated that these legends and stories are all that remain of a recollection of the time when the Star People in their sky-crafts regularly came out of the western skies, especially during the "The War in the Heavens," and directed light beam weaponry (which resembled lightning) at colonies of their enemies. The great rumbling heard on Earth was the sound of their sky-craft as it approached and then passed overhead.

Another legend tells of a time when a serious quarrel arose between the beings called "Underwater Panthers" and some other beings called "Thunderbirds." The Underwater Panthers have a "den" in the depths of what is now called Devils Lake, in Sauk County, Wisconsin. The legend states that the Thunderbirds hurled "thunderbolt arrows" into the waters and onto the bluffs surrounding the lake. The Underwater Panthers "threw" great "rocks" upward from beneath the waters of the lake in an attempt to hit the Thunderbirds. A terrible battle continued for days. The tumbled-down and cracked rocky surfaces of the bluffs surrounding the lake are evidence of the great struggle. Finally the Thunderbirds were victorious and soon afterward they flew away to their "nests" in the northwest.

No American Indians would approach that lake for a long time. It is said that the Underwater Panthers were not all killed during the war and that some still live in Devils Lake to this day.

Later, American Indian peoples living in the area not too far from the lake, as well as travelers from afar, had a custom of making Kinnikinnick offerings to the "spirits" of the lake. They would deposit their offerings on boulders along the shore or place them on the surface of the water.

The Elders said that stories such as the one told here were the memory remnants of actual accounts of the fiery sky battles between two very different forms of Star People.

It was believed, according to the Elders, that at one time in the ancient past the Thunderbirds had a huge "nest" in the then-mountainous area around Lake Nipigon, northeast of Duluth, Minnesota, in Michigan's Upper Peninsula. They said that, according to legend, large blankets of clouds always covered the "nest." The native people in the region considered this "nest" area to be a sacred place and did not go there. Consequently, they did not really know what the Star People were doing in that area.

According to the Elders, some years before the Europeans arrived, the blanket of clouds began to lift and move away. At that time the Thunderbirds destroyed every trace of the place, pretty nearly leveling a large portion of the mountains in the process.

The Elders also said that, in ancient times, there were humanoid-appearing "Water Beings" in the Great Lakes area. They also were related to the Star People. These Water Beings sometimes approached humans, and instructed them by communicating through telepathy. Their craft were very different from the water craft of the humans. Long ago, human ancestors erected offering rocks for the Water Beings at various areas along the shorelines. One such offering rock lies at the northwest point of Medicine Island. The Elders said that, since the coming of the Europeans, the Water Beings almost never appear, however, it is believed that these beings are still around. Once in a while if someone is caught in a storm and reverently makes an offering to them, the Water Beings may help calm the waters.

The Elders said that the original humanoid was lowered to Earth from a starship. They said that the initial Earth home for "Original Man" was a large land mass that was located east of Central and South America in the mid-Atlantic—a land mass that obviously is not there now. They explained that the area was destroyed when the great flood completely covered the land mass. It is still under the waters of the mid-Atlantic. (This area, incidentally, is part of the area destroyed due to the misuse of power.) The Elders said that the word "Anishinabe" originally referred to "the people who came from beyond where the sun rises."

Original Man, humanoid but not human, is a hybrid who embodies some of the most negative qualities of his human ancestors and some of the most positive qualities of his Star People relatives. Often considered a nefarious character, he is in actuality neither all bad nor all good, but simply a reflection of both. Because of this, he is what some scholars refer to as the culture hero or trickster.

Original Man spent quite a good deal of time visiting and learning from Nokomis while she was still living on the Earth. Later Nokomis, like her daughter, left the Earth, to live elsewhere in the universe.

Whenever the moon is visible in the skies at night, and especially when it is full, humans are to remember Nokomis and her special role as "grandmother" to all human beings.

Starpeople: From a Scientific Perspective

In the branch of metaphysics called ontology, the nature of "being" in the universe is approached by asking questions about the types of beings that exist, or the nature of these beings, or whether any possible relationship could exist between such beings.

American Indians of the Great Lakes region answered these ontological questions in their ancient legends and stories. It is clear that, like native peoples throughout the world, they believed in the existence of other-world beings. Their stories also suggest that communication and interaction between humans and those other-world beings occurred. There must be some reason beyond mere coincidence that explains why the commonly held metaphysical belief occurs in so many ancient human societies.

Contemporary theoretical scientists, especially those involved in planning the exploration of space, ask the same ontological questions as the ancient tribal people, but from another perspective.

In July of 1989 the National Aeronautics and Space Administration (NASA) awarded a three-year grant to the School of Architecture and Urban Planning at the University of Wisconsin—Milwaukee to support research related to designing work-space and living quarters for planned colonies on the Moon and Mars. Other United States universities are also involved in research concerning the exploration of space and proposed expeditions to Mars. If we possess the technology on Earth to make other planets suitable for habitation by humans, then is it not possible that other beings may have done the same on Earth? Could it be that the existence of the Star People is theoretically possible from the perspective of modern scientific knowledge? Why else would the space probe Voyager, which will eventually leave our solar system, contain pictures of what we look like, a map of our solar system, Earth's position in it, and technological information in formulae, if those people working on the project did not believe that the existence of other life-forms were possible?

In 1981, Professor Francis Crick, who won the Nobel Prize for Physiology in 1962 for his work on DNA, wrote *Life Itself: Its Origin and Nature*, in which he describes our vast universe containing at least ten billion galaxies, each of which has stars that resemble our star, the Sun. In our galaxy alone, the Milky Way galaxy, there are somewhere near one hundred billion stars.

Based on these observations, Crick asserts that it is logical to assume that many stars in the universe have planets orbiting them and that a significant number of those planets have the conditions necessary for sustaining some form of life.

Crick postulates that "thinking life" must have evolved on a percentage of the planets. He states that it is logical to assume that scientific and technological developments may also have taken place on some of these planets.

It also follows that some of these developments may be far beyond anything humans have accomplished to date, and may include the ability to travel to neighboring planets, or to nearby star systems, and possibly into other galaxies.

Crick speculates that peoples from one or more of these worlds may have known that their own civilizations were, for whatever reason, doomed. Perhaps they had found that a neighboring star was set on a collision course with theirs or that their own sun was going to become a red giant that would eventually engulf their planet.

He states that it is possible, perhaps even likely, that inter-space flight did occur wherein highly advanced beings traveled and colonized other planets in other star systems, and that upon arriving at these new planets, they may have had to carry out extensive genetic engineering in order to exist, given the environmental peculiarities of the planet.

He states that we cannot rule out the possibility that large spacecraft (mother-craft) maintained, and perhaps still maintain, orbits in places where they would be difficult to detect—near Neptune, for instance, or in the asteroid belt between Jupiter and Mars. These beings have probably made, perhaps still make, forays in smaller crafts to planets in our solar system, including Earth, to obtain needed minerals and other raw materials, as well as to observe.

Crick concluded that it may be "dangerous" for us to assume that we are "alone" in the universe and that we should be cautious about our actions on this planet lest we run the risk of being perceived as about to "contaminate the galaxy" through environmental and technological abuse.

It is interesting to note that, just as the Elders pointed out, Professor Crick also stresses the need for balance in our daily actions within the context of the other beings who share our universe. Events on our planet have repercussions in the universe that 20th century scientists are only beginning to understand. If there were nothing else than this to learn from the stories told here, they would still be invaluable. Let us hope that the Star People will not have to hold us accountable when one day our planet ends in destruction because we refused to believe people like Crick and because we ignored the ancient stories of the Elders.[10]

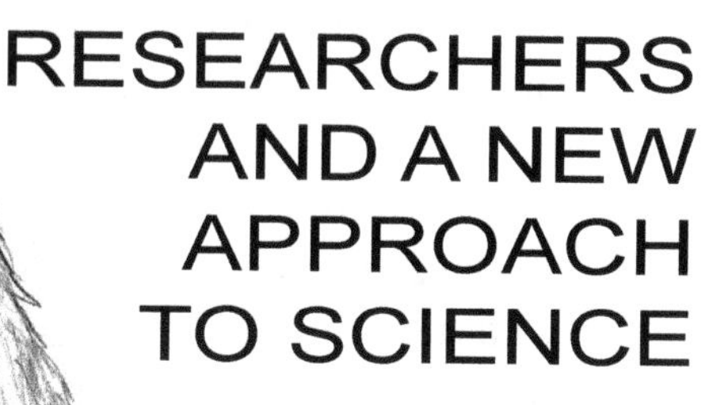

CHAPTER 8

RESEARCHERS AND A NEW APPROACH TO SCIENCE

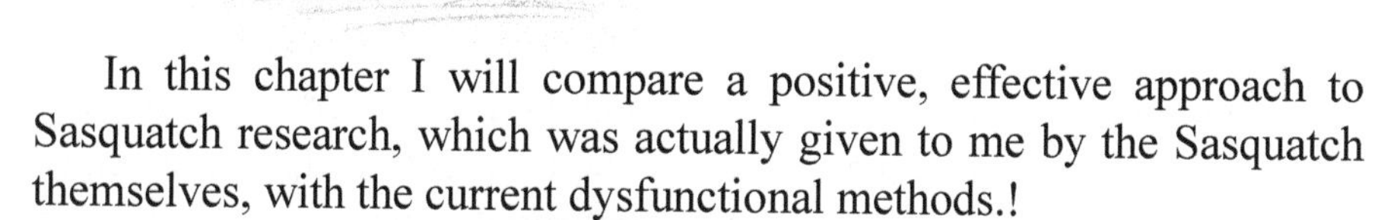

In this chapter I will compare a positive, effective approach to Sasquatch research, which was actually given to me by the Sasquatch themselves, with the current dysfunctional methods.!

Those who insist on having a dead body have missed the essence of the Sasquatch phenomenon: the fact that they are a more evolved humanoid species than we are, and that their psychic abilities, which they currently use to elude us, *could* be used to help us evolve!

With all the helicopters and technology and searching, where is the evidence? There is none and won't be any! Why? Because the psychic Sasquatch *is in control*—not misinformed man. Furthermore, if a person kills a Sasquatch, then it can no longer show us who and what it is. When dead, it merely becomes a biologic mass of flesh. Its "creatureness" is stolen and it cannot share its wisdom and living mastery of quantum physics, that is, psychic phenomena.

Further, if Sasquatch researchers would study the UFO phenomenon they would immediately notice a perfect parallel between ETs and the Sasquatch people—appearance/disappearance, mental telepathy, psychic healing powers, advanced intelligence, and spiritual awareness. Why? Because the giants *are* ETs.

Of what use are trail-cams, heat-sensing devices, and all the other overrated technologies when the only link the Sasquatch will allow is with the heart instead of the head? Their sincere camaraderie with me and others who approach them with unconditional love, not fear, is proof enough for witnesses who experience them. Capturing their friendship is the real success!

Those mainstream researchers have automatically assigned a subhuman status to the Sasquatch, similar to what Native Amerindians, Blacks, and Asians endured when Europeans/Americans encountered them. That prejudice has now been *transferred* to the Sasquatch people, who are still being hunted and harassed by researchers who think they are animals. Researchers are also motivated by money, status, egotism, and a wish to be accepted by the larger group—ultimately, mainstream science itself—rather than to objectively follow the clues to where this phenomenon leads. All the anomalous data being discarded *is* the pot of gold wherein the real truth lies! And the so-called weird psychic stuff being repeatedly reported is really a part of mainstream science disguised as quantum physics!

The ongoing problem with Sasquatch researchers is that they believe that their subjective actions, when collecting data, are really hard-core science. But how can cryptoanthropology progress if they are tainting the research by not following scientific protocol? Objective data collecting means documenting exactly what witnesses are reporting—not ignoring it.

Archeology researcher Michael Cremo calls it "knowledge filtering," wherein new data that do not subjectively please pet theories are deliberately hidden or conveniently ignored.[1] They make historic reference to the Sasquatch and previously-described events during their quest, but have never spent any time talking with a Sasquatch to find out how he/she thinks as a unique type of human! This is a serious problem, since the scientific establishment as a whole, the public, and the world start to believe in a false view of Universal Reality. This data is then used to concoct a theory that further perverts the real truth, which ultimately is accepted as the norm. As a result, none of them have made any progress over the last half century of searching.

The necessity of a change in Sasquatch research is succinctly stated in Ida Kannenberg and Lee Trippett's 2009 book *My Brother Is a Hairy Man*. An ET told Ida:

> The reason all those researchers have not been able to find patterns in Bigfoot's activities is because they have been collecting and studying data based on him being a curious animal with a big brain only, and carefully listing all the characteristics that would make him so. Until they consider his human traits, and study his higher mind and soul, they will never prove him anything. They are afraid to discover his humanness. It raises too many questions about their own. That is why your and Mr. Trippett's book is important. If Bigfoot is to be saved from extinction these facts must be brought out. His extinction would only presage your own.[2]

Veteran researchers Lee and Marlys Trippett followed the evidence that led to a psychic Sasquatch.

In Sasquatch research, most researchers, starting at the top with the veterans, have subjectively *thrown away* or ignored or not followed up on vital data over the last fifty years because they didn't believe the subjective data, i.e., what actual witnesses were telling them. Roughly 99% of researchers now follow suit! In other words, in an effort to *look* scientific, they ignore and/or alter data because it's important for them to be accepted by their peers who, like themselves, have never had a one-on-one encounter with the Sasquatch people.

Science teacher Thom Powell wrote an excellent book called *The Locals* using research data that had been *subjectively* discarded by California-based Bigfoot Field Research Organization (BFRO) for which he was a curator. In the book, he states that it "is the world's largest organization that *objectively* investigates the Bigfoot Phenomenon."[3] He said the data was excluded because it included mental telepathy and UFO information from interviewees! Though he has never encountered a Sasquatch nor experienced mental telepathy, he writes a very compelling story. My question is: Why was valuable knowledge being thrown away?

In fact, *the anomalous data is the "missing link" to the entire phenomenon.* It is the *most* important data because it includes direct communication with research subjects. To repeat an old adage by William

Researcher Thom Powell with author debating the Sasquatch issues

James, "It takes only one white crow to prove all crows are not black." I have documented 187 "white crows"—people from all walks of life who make extraordinary and baffling claims about their experiences with the psychic Sasquatch.

Instead of doing field research trying to contact the Sasquatch people directly, those who do a majority of their research surfing the Internet typically bash contactees. For example, cryptozoologist Loren Coleman wrote a blog on his website in July 2010 targeting "contactees" in a negative way. No one has a right to tell contactees that they are wrong in what they experienced. As the Sasquatch have said, "Stand in your truth, that is all we ask."[4]

Those researchers who are on the right track, according to the Sasquatch themselves, are also subject to attack and misrepresentation.

One scholarly person who should be commended for his fine academic work in the field of Sasquatchery is Professor Jeffrey Meldrum, PhD at Idaho State University—Pocatello. Numerous newspapers carried a story about Meldrum being harshly criticized by fellow scientists at the university for dabbling in "pseudoscience." Actually the thirty professors who petitioned the dean of arts and science against Meldrum are violating their own scientific philosophy by not exploring new areas of science. Kudos to Dr. Meldrum for his courage against such odds.

I myself have had the mixed blessing of being mentioned and featured in some twenty-three books over the last decade. One of those who got it wrong was author/paranormalist Rupert Matthews in his 2008 book, *Sasquatch: True-Life Encounters With Legendary Ape-Men*. Matthews incorrectly states, "Lapseritis suggested that these were not encounters with a real animal at all, but with the paranormal."[5]

I would not have said that. While lecturing at the 1998 International Sasquatch Symposium at the University of British Columbia, an audience member asked me, "Are they paranormal or physical? They can't be both!" My reply was, "That's the big contradiction with this phenomenon—they *are* both!" At times, I have seen just an outline of their body with no definable features. I have also observed them as living apparitions with all features clearly visible. And, I have encountered them in the physical state seven times. They are interdimensional, yet physical, with the ability to be paraphysical—that's the reality. Scientifically, quantum physics offers some understanding of this.

And then, Matthews stated, "Jack ('Kewaunee') Lapseritis has since dropped out of the Sasquatch scene, but his ideas have continued to circulate."[6] Nothing could be further from the truth, and there is no valid excuse for misinforming the public about my status. In fact, I am one of the most active field researchers ever. Not only do I explore new areas where Sasquatch abound, but, at times, interact with Sasquatch weekly at my isolated cabin in the forest. My address and telephone number are all over the Internet and I welcome anyone to call or write me with their questions. People contact me all the time. After a half century in the field, I continue to enjoy searching new horizons to have more new experiences with the psychic Sasquatch.

Another point of vulnerability, when ninety-nine percent of researchers lack direct personal experience, is the real risk of fraud. This is not easy research to collect or to write about.

Researcher Autumn Williams wrote a book called *Enoch: A Bigfoot Story* (2010) about a Florida man named Mike who claimed to have had a close association with a Sasquatch person there. It appears that Autumn Williams did her best to convey the truth as she understood it, and conducted a thorough analysis of Mike's claims—except to visit him in Florida to get a feel for his veracity. But from the beginning, I found numerous contradictions, based on my personal experience and on my many interviews with other contactees.

For example, Mike claimed to have followed the giant, witnessing the killing of several raccoons during the daytime, which doesn't make sense because raccoons are nocturnal. And how could a Sasquatch consume catfish as described without their sharp spines stabbing him in the throat. "Hornpout" is another name for catfish, because they have three sharp spines on their fins that punctured me more than once as a boy. Also, Mike carried a .44 magnum and shotgun, but Sasquatch would never get near a person carrying a gun, because they know guns can kill! There are more inconsistencies. For example, he described the Sasquatch as "dog-paddling" when swimming, when on every other occasion, Sasquatch have been observed to do the breast stroke. And laughably, Mike says that the Sasquatch's powerful odor is due rotten meat getting caught in the hair on their chests—the exact same explanation given by known hoaxer Ray Wallace over the years. My understanding is that they have a musk-like gland that is activated when angry or startled—similar to a skunk's.

Then, in December 2010, a few months after the book was released, the Mike character made a public statement that the entire material for the book was fictitious—he made it up!

Great discernment is necessary, but sometimes that's not even enough! In 1986, after sixteen months of weekly interviews and collecting of data in the forest, I discovered that I myself was being conned by an old man in Oregon who wasted my time fabricating tales of Sasquatch and ETs. I am convinced he had initial encounters, but I suspect that when the contact stopped, his need for attention drove him to concoct wild stories. After I noticed a series of inconsistencies from what he initially told me, I politely thanked him and left. Then I discovered he had stolen some of my research data—four sketches I had paid an artist to produce. Then he stole my *identity*, cleverly reversing the situation by saying I had stolen from him. In trying to resolve this horror, I visited his hometown, talked with his wife and adult children—and I discovered that this man had victimized many others, too! The giants and ETs do not get involved with deceitful people. Unfortunately, because his fraudulent information was posted on the Internet, it's still out there, continuing to hoax the public.

Frauds often hurt the researchers as well as the field of cryptoanthropology and cryptozoology. Science is already skeptical of anomalous data, and any deception puts a tremendous strain on the credibility factor in this field of study.

Even I, as a researcher with academic credentials and as a Sasquatch/ET contactee, cannot produce proof of what I am stating, though I am not trying to prove anything. The psychic behavioral patterns and telepathy of the creatures—that I and the other 187 witnesses I have observed—are considered subjective because not everyone experiences them, even though those who do are consistent in their experiences. Two physicians, a college professor, an attorney, and people from all walks of life have objectified my reality when making these anomalous claims. So there is significant substantiation in my work. Plus, a majority of contactees do not want their names to go public, which means they are not seeking notoriety. The beauty and privilege of experiencing communication in a Sasquatch's presence far exceeds any need to obtain physical proof, as I personally discovered in 1979 when first contacted by the Sasquatch and ETs. The quantum-skilled Sasquatch have proven to me and others that they exist.

In 1981 I abandoned my efforts to acquire physical proof, knowing it was fruitless. My personal interactions with these beings became more meaningful and profound. The encounters are still ongoing. Science and researchers have never been up against such a clever being before in all of human history. The American Indians knew of the elusiveness and unusual abilities of the giants. I feel it is vital that researchers seriously rethink their present field methodology, which harasses the Sasquatch and often drives it away or puts it on the defensive.

My field approach has been successful for over three decades now. Most years I conduct two to four field expeditions in order to contact and interact with different tribes.

Ninety-nine percent of the time I camp alone with no fear, no gun, and no camera—just "free-flow" and a feeling of love for every living thing in the forest. "Talking" to birds, deer, trees, and flowers helps one to become attuned. Being a healer, I carry a huge amount of compassion and faith in our Creator. Successful research is literally the way of the shaman!

Meditation is grounding and helps one push aside the cumbersome intellect while transcending the ego. Henry David Thoreau said, "Our truest life is when we are in our dream awake," and that is what I seek:

Author's shoe beside a track of an Ancient One in southeast Oklahoma, March 2006

oneness with the giants—a yearning to live with them on their terms and learn a totally different culture and life. It's like stepping into a science fiction movie, except it is all real.

At times, I admit, I feel apprehensive while hearing the weight of the giants walking toward me in the darkness of the forest as I sit on a fallen log. But I merely remind myself that God and love are synonymous and that in my heart I am their friend. It takes practice to control one's thinking and emotions, yet I like the challenge of a new way of being. The Chinese philosopher Lao Tzu (c. 604-531 BC) said, "There is no greater illusion than fear...Whoever can see through all fear will always be safe."

And I have weekly encounters where I live, in the forest in the foothills of the Cascades several miles from Seattle, Washington. Occasionally, they knock on my door when I'm up at two or three o'clock in the morning reading or writing, to let me know they are there. They bang on the side of the cabin when I have a visitor. The Sasquatch interdimensionally come inside and deliberately wake me up at night with a deep, loud heavy breathing to let me know they are watching over me. The beings give me sage advice when I'm sick. In the daytime, they

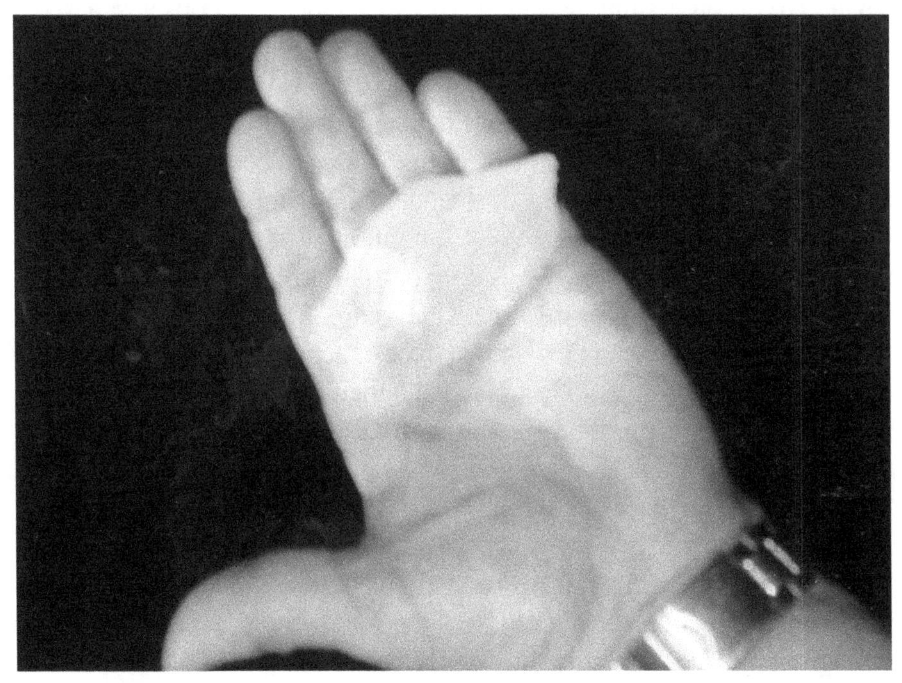

A crystal left for the author when food was left for the giants

have stood at the edge of the forest to talk telepathically to several of my house guests. This has been going on for the eight years that I have been living here. It's wonderful! This is real and valuable research.

Frequently, guests who spend a few days at my cabin have friendly ET encounters as well, even powerful healing if they are ill, as I have detailed elsewhere in this book.

In July 2006, here at my cabin, Mary Rau visited me. We had communicated but had not yet met. She set up her tent at the edge of the forest and I asked the giants to visit her. The first night they walked up to the tent to let her know they were there. The following day at 2:10 p.m. while Mary was alone reading a book on the porch, a ten-foot-tall Sasquatch named Elin appeared some 60 to 70 feet in front of her standing in the forest. She reported that it was not scary, but pleasant looking. Mary initiated telepathic communication and he spoke to her for a few minutes in a most intelligent manner. Since then, she has had more in-depth psychic encounters in the Sierra Nevada Mountains of California.

There was a gathering of several researchers at my cabin on June 30, 2010. When one of the people walked out to get something in her car at 7 p.m. on a sunny evening, she was shocked to observe a six-foot-tall Sasquatch run from the "inside" of my cabin (where the rest of us were socializing) straight through the solid door, through the solid railing, leap onto the tiny patch of lawn, and run into the thick forest! I had told the other researchers earlier that we were being monitored by a Sasquatch *in my cabin.* It was true. The incredible thing the person noticed that caught her immediate attention was that she could see right through the creature! It was only partially solid. (I have seen this, literally, a few hundred times over the years. Their eyes moved and their skin wrinkled when they changed their facial expressions; yet they were living "apparitions.") Later, two 17-inch Sasquatch tracks, which would have been from an 8–10-foot adult, were found in loose dirt at the edge of the woods. The creatures were attracted to our friendly get-together and enjoyed listening in.

Once, I had the privilege of meeting a very kind gentleman named Mack who lives in Arizona. He knew little about the Sasquatch people, yet had an attraction for them. He asked for my help, so I marked on a map where he should go and told him about a group of Sasquatch working with friendly Starpeople. Mack was cautious, but proceeded on his own to this wilderness place. He observed real spaceships, and a female Sasquatch reached into his tent and held his hand. He overcame his fear in exchange for love and trust. Mack also encountered a ten-foot, white-haired giant staring at him from a distance. They knew he had no guns or cameras, as he had followed my instructions. One night, Mack saw two UFOs land behind a large mesa, pursued by two black, unmarked helicopters shinning spotlights around the gigantic rock formation. Another time, he accidentally backed his truck into a ditch with the back wheels suspended in mid-air. In front of him on the opposite side of the road was a dangerous cliff! This occurred just before sundown. He decided to telepath to any Sasquatch who might be in the area for help, just as I had taught him. In less than fifteen minutes a Sasquatch picked up the back end of the truck and placed the wheels onto the gravel road. Mack was late for supper, but at least he arrived home safely, thanks to a caring, helpful giant. This happened in the spring of 2010.

When the gentle and kind Mack visited me at my forest cabin in Washington State for a week, he was happy to pitch his tent just outside my door. One night a Sasquatch reached into his open tent and carefully moved each of his legs to one side. He simply thanked the giant for trusting him in that way. It was Mack's childlike (not childish) affection and

Mack visiting the author at his cabin in Washington State, where he had a Sasquatch experience

innocent demeanor that attracted the Sasquatch. Mack merely wanted communion and nothing more. Suzanne Scurlock-Durana states, "The latest brain research shows that when love and compassion are present, the brain lights up and operates to its fullest capacity."[7]

I know that the Sasquatch would welcome more social interaction with other people like Mack who visit their forest domain. When veteran researchers learn to take on Mack's attitude, then there would be no need to look for Sasquatch, because the giants would go out of their way to find them! This research is of the *heart*—not of the mind and intellect.

Author exploring Drunken Charlie Lake in the Washington Cascades where occasional sightings have occurred

Contactee Kathleen Jones was told the importance of this many times. There is a whole new world of excitement and knowledge waiting for those who can let go of fear and thoughts of violence. This is the vital message that will resolve the ongoing problem.

Helping other experiencers make meaningful contact has been most rewarding. In 2009 I began training specific people and do so *free*! I warn people to beware of those who charge a fee to tour the forest looking for Sasquatch. Some have put my name on their websites as if I were promoting their business, but I would *never* endorse anyone who makes such claims.

I have spent a great deal of time on the phone these last ten years counseling and being supportive of other contactees. They need to know that encountering a psychic Sasquatch is a special experience to be embraced. Contactees need not be concerned what those who are ignorant say about them. Most of the nay-sayers would be less sardonic if a Sasquatch telepathically spoke to *them*, and I wish the giants would! In some way contactees are special people who have the right stuff and are less ego-involved than the others. Once one knows that mental telepathy is real and valid when communicating with a sentient nature being, then the contactee has no doubt that there is a psychic Sasquatch.

Although I am also an experiencer, I only tell interviewees about my own anomalous encounters after the questioning is complete. That way, percipients are assured that their psi experiences were indeed real, which has a cathartic effect on each of them. And their experiences also objectify my own reality, since they parallel my own.

We must go back to examine the fact that the forest giants are incredible mind readers—the ultimate spies, if you will, trained to be clandestine by a secret forest society protecting themselves from the dangers of superficial man. There is nothing unusual about radar in bats; sonar (echo location) in whales and dolphins; a form of electromagnetism in birds; vibrational receptors in the jawbone of crocodilians; or a shark's ability to smell blood in water almost a mile away. We accept without question these fascinating attributes in these animals, so why is it so difficult to accept a psychic Sasquatch who uses anomalous powers to survive?

Often contactees will make the best researchers, especially when the Sasquatch live in their backyards. This allows the giants to "listen in," and in so doing, get to know the experiencer better. The following contact from my files illustrates a fairly typical situation that allows behavioral patterns to be easily documented. In March 2010 I received a telephone call from a woman in Colorado. She was experiencing a young Sasquatch that was trying to enter her house at night through the back door. He had broken part of the storm door and made nightly noises for nearly two weeks, which frightened the woman immensely. She had contacted researchers who advised her to erect outdoor cameras and another group arrived giving similar instructions. The advice I gave her over a two-hour phone call was to put food items out, not cameras; plus I taught her mental telepathy.

During our conversation two Sasquatch apparitions appeared next to my desk, as they were monitoring the fearful woman's plea for help. I asked the beings if they would go to Colorado from Washington State to ask the inquisitive Sasquatch to please stop scaring this woman. They agreed! Within one minute she could feel "someone" watching her in her house. As we talked she eventually became calm, releasing a lot of anxiety. Three weeks later, she called me again to report that peace had come over her residence starting the night I sent the two Sasquatch to assist in resolving the matter. From time to time a small rock would hit the side of the house, sometimes hitting the window. She would telepath, asking them not to hit the glass anymore and her request was always granted.

Then one late afternoon after snow had fallen, she noticed two neighbor kids running across her back lawn, which was adjacent to the national forest. But a second look startled the witness when she realized they were two young Sasquatch! Their "fluid" movement when running was unlike anything she had ever seen. They were both dark brown from head to toe. The next morning there were creature tracks all over her backyard in the snow.

I continued to encourage the woman to telepath love and kind thoughts. She told the adults they are welcome to come into her home and to help themselves to any food they wanted. It was explained to the experiencer that the forest beings are interdimensional and could enter a house easily, as they frequently do at my cabin. One night she heard a noise in the living room. As she stepped into the room an 8–10-foot, shadowy Sasquatch was just walking through a wall! This is normal for them; paranormal for us. The giants enjoy hanging around people who invite them in and leave food without deception. This is the theme for success. As Dr. Wayne Dyer said, "When we change the way we look at things, the things we look at change."

It took nine months for the woman to completely exchange her fear for love and trust, at which point the creatures would telepath back to her. It takes longer for some people than others. She commented that the telepathy is like being in a dream—at first it didn't seem real. The more a person lets go of the layers of anxiety, the clearer the communication will be. Patience is necessary. Now she misses them when they are out of the area. The experience has profoundly changed her life!

There is much to learn from the Sasquatch people, if only investigators would shift their thinking and trust in a more gentle, congenial field approach. In every field of endeavor there is competition, professional jealousy, harsh criticism, heavy debate, and even character assassination. The Sasquatch people would feel good if researchers would get out of their heads and into their hearts and act respectfully to other researchers, to everyone, and every living thing. This is simple and basic.

Mother Theresa sums it up best:

People are often unreasonable, illogical, and self-centered, forgive them anyway;

If you are kind, people may accuse you of selfish ulterior motives, be kind anyway;

If you are successful you will win some false friends and some true enemies, succeed anyway;

If you are honest and frank, people may cheat you, be honest and frank anyway;

What you spend years building, someone may destroy overnight, build anyway;

If you find serenity and happiness, they may be jealous, be happy anyway;

The good you do today people will often forget tomorrow, do good anyway;

Give the world the best you have, and it may just never be enough, give the world the best you have anyway;

You see…in the final analysis, it's all between you and God…it was never between you and them anyway.

CHAPTER 9

THE FUTURE OF SASQUATCHERY

Peter Miele and his wife were visiting the Kingdom of Nepal in the Himalayan Mountains, where the 29,028 foot high Mount Everest touches the sky. It was 1974. The two adventurers were trekking outside of Kathmandu to explore the countryside—particularly a place called the Monkey Temple. Previously, in June 1968, I had spent two and a half weeks there investigating Yeti sightings and, coincidentally, also visited the Monkey Temple.

When the pair arrived at the temple, they began climbing up very long, steep stone steps leading high into the monastery. Part way up, a big monkey jumped down from a stone wall and grabbed the wife's handbag, gripping it tightly. She was terrified! Peter, in the lead, was several steps higher. He quickly walked back down to his wife's defense. Peter is an animal lover, but was afraid the monkey might bite, so in a panic, he did the only thing he could think of, which was to kick it. He kicked it harder than he had anticipated and it twirled in the air, landing on one of the lower stone tiers. Immediately it became vicious, snarling at the frightened couple. Peter stood his ground by taking a large knife from its sheath in the event the monkey decided to attack. They slowly backed away to safety, then continued to climb upward toward the temple.

Once inside they spent an hour and a half talking to the monks. Then Peter and his wife decided to browse around the temple complex. They noticed a sunny meadow, half the size of a football field, with an idyllic forest set against the mountainside. The couple wanted to rest amidst the greenery. They found a stairwell leading down to the field, then they casually walked to the edge of the forest.

No sooner had Peter and his wife reached the trees when several large bushes began shaking violently, which was enough for them to have an immediate concern. This intense shaking activity continued along the row of trees inside the forest and Peter began to think it was a tiger. They were scared! They stood waiting with trepidation, wondering what animal might emerge from the dense forest.

Suddenly a large hairy primate leaped out onto the open meadow! They were stunned. The creature stood erect with a threatening posture, staring menacingly at them. To add to their tension and fear, four more bipedal hairy beings stepped out from behind the first one. Now there were five! What instantly caught Peter's attention was how vastly different the latter four man-creatures were from the first. The ominous-looking one had short, black hair with a human stance, arms in proportion to its body exactly like a human, plus a human-type nose and eyes. Yet his overall features resembled a huge monkey on two legs. It was a monkey-man, not ape-like in any way. The other four had longish, light red-black hair, arms extending down to their knees, were slightly hunched, facial features somewhat ape-like, and their bipedal gait was totally different than the monkey-man's. None of the five males had a sagittal crest or peakedness to their heads. All had round heads similar to *Homo sapiens sapiens*. Plus they were all the same size—five feet tall. Then, to trigger a higher anxiety, Peter said, the five of them sprinted toward him with super-human speed unlike any living thing he had ever seen before! Their speed was unnatural.

Peter Miele exploring isolated temples in the jungles of Cambodia, where he discovered a stone mural of a giant hairy being holding a sword, fighting a group of warriors

Monkey-man, the apparent leader, stuck his face in Peter's face some two inches away. The creature was so close that Peter could feel its hot breath against his face. The being stared ominously with human eyes into his. Then it sniffed his body all over without ever touching him. To add to Peter's horror at such a confrontation, the simian figure had "fangs"—two protruding three-inch long canine teeth!

Looking around, he became aware that the other four figures had formed a tight circle surrounding them. These four were a mere two feet from Peter, as his fearful wife huddled close behind his back. There was more intimidation from the black fellow who was growling in Peter's face, displaying a powerful looking arm and hand, which had three inch long pointed fingernails resembling a claw! His thoughts were that this creature could rip him to pieces if it so wanted. As the threatening behavior continued, Peter realized that the reason for the creatures conduct had been instigated two hours before when he kicked the monkey. He said he could read the intimidator's body language, and psychically he was saying, "Pick on someone your own size. You want to try that with me and you will lose!" Peter noticed that the man-creature's forearms were as big around as his thighs.

Finally, there was a bit of relief when the being turned and walked away. The other anthropoids instantly followed. But, when they were some 15 to 20 feet away, the five creatures turned around and charged him again, and the confrontation started all over again with the monkey-man growling and showing Peter his lethal looking hands. Once again they walked away. After some 30 to 35 feet, the leader turned around and with tremendous speed ran up to him, starting the grim ritual all over again. Finally, when the group reached the far end of the field they ran back one last time—apparently to reiterate that it's wrong to harm an animal, especially if it's smaller and defenseless. Perhaps, like the Sasquatch and Ancient Ones, these beings were protectors of Mother Earth and her "children."

This amazing account, I am convinced, is authentic. Peter told me during our interview that he went to the department of anthropology at a nearby university after he returned home, and was dismayed by the indifference of the professors. Because it was anecdotal, the personal experience he had was considered invalid information to scientists. Indirectly he felt they were labeling him a liar. Even though Peter Miele has two university degrees, has traveled to 87 countries during his life, is articulate and very observant, he told me he was discouraged when trying to share this extremely rare encounter.

From his anatomical description, none of the creatures appeared to be Yetis in the classical sense. A Yeti is classically described as seven to ten feet tall, with blonde-grey hair, arms down to the knees, and a bullet-shaped head. Because of their benign actions, and because they did not harm the couple in any way, I feel Peter was characterizing humanoid beings, *not* animals. The leader was very calculating in its threatening behavior; but if they had been five gorillas or chimpanzees in such close proximity, the two people would have been torn apart. Why didn't the bipeds simply kill them? The Sasquatch and Starpeople have spoken to me about their belief in karma—that is, that we all are spiritually responsible for our actions. To kill a person is heavy karma. They were not baneful. The Sasquatch and Ancient Ones said there are many races of their people worldwide, and these are an example of them. It also means there are "at least" three separate races of humanoid beings in the Himalayas, *not* just the Yeti. Granted, the Yeti have displayed aggression and fury during many close encounters. My response is that humankind, like locust, is presently invading the last hinterlands on the globe. With our exploitations, we are pushing wildlife and indigenous peoples to the very brink of extinction. The Yeti, Sasquatch, Ancient Ones, et al., are acutely aware of this fact. In Brazil's Mato Grosso, intruders disappear all the time, never to be seen again. A hidden Indian tribe may be responsible, knowing that it's at risk of vanishing, decimated by greedy outsiders. I feel the Yeti and other humanoids are merely defending their turf in their struggle for self-preservation. In our increasingly overcrowded world, their very survival is at stake!

The future of Sasquatch-types extends to Latin America, where the largest uncharted region of the world exists. The following account gives one an idea of what might be living there, if only we humans will survive long enough to discover it.

Archeologist Pino Turolla, in his fascinating book *Beyond the Andes* (1970), tells of a personal encounter with "el mono grande" (big monkey) in the jungle of Venezuela. This greatly interested me because I have visited Venezuela twice, the second time camping near Angel Falls for five days, which is at the beginning of the remote Guyana

Highlands. Angel Falls is 3,212 feet high, over twice the height of the Empire State Building!

Turolla was exploring a section of the rainforest with a jungle guide plus two Indian helpers. The area was extremely remote, and undefined on a map. He relates that the undergrowth was so thick that it reduced his ability to see to only a few feet. Eventually the dense foliage opened to some degree into a canyon where his guide Antonio had previously had a lethal encounter with a troop of six-foot tall "monkeys." A few years before, Antonio and his two sons had entered the canyon and were attacked by three bipedal primates wielding clubs. He and his sons ran from them, but his youngest son fell and was clubbed to death by the creatures. Turolla said that Antonio swore to the truth of the story, expressing a lot of fear and emotions as he related the experience. The party of four became very apprehensive.

> It was getting toward late afternoon when suddenly we heard a howl, very loud, coming from somewhere in the thick vegetation. The Indian froze. The howl was as loud as the roar of a jaguar, but it was higher and more shrill in pitch. It reverberated through the forest, encircling us as if it came from all directions. Something was moving, crashing powerfully through the underbrush.[1]

This occurrence bothered Turolla for a long time. He said he had heard about Bigfoot, but was unsure of its reality because of a lack of proof. Now, here *he* was, a witness to the hairy folks' reality, but without any evidence of its existence.

Incredibly, two years later in the jungles of Ecuador he had another hair-raising encounter. Turolla and Oswaldo, his guide, were seeking a certain cave by the Rio Paute to look for artifacts. He uses the Spanish word "cueva" for cave in the book. Though I have spent time in Ecuador, I am not familiar with the eastern region he explored. My time there was spent visiting the coastal area, as well as a one-month stay while island-hopping in the Galapagos Islands in 1970.

Turolla and Oswaldo were forging their way through the jungle on horseback in a heavy downpour. They looked forward to finding the entrance to the cueva to get out of the rain and to possibly make camp inside it. Eventually they found the cave. Both men cleared the entrance with their machetes. They climbed in, took out their flashlights and found themselves in a dark passageway that led further into the mountain. As the pair crawled over rocks, boulders and debris they noticed "a stinking smell." The smell grew more pungent as they cautiously worked their way down a large tunnel some five meters high and managed to walk 50 to 60 meters further in.

> Then, suddenly, the tunnel opened up into what we sensed to be a huge chamber. We hesitated, hands on our guns, wondering whether to enter, our backs pressed against the safety of the wall. We stood there beaming our lights in all directions, unable to perceive the dimensions of this enormous black hole. There was no sound. After a few moments, we relaxed our nerves by lighting cigarettes. This curious hiatus of darkness and silence must have lasted less than five minutes when it happened: an incredible, blood-curdling, howling roar, louder than the sound any man could make, coming from the depth of the cave. Then a crescendo of howls and paralyzing screams, higher in pitch and more intense. We stood there rooted with fear, our hearts hammering, staring in horror at the barrier of blackness, our lights making erratic patterns on the floor.
>
> Oswaldo grabbed my arm and I felt a penetrating pain in my muscle. I tried to react, but then, amid the terrifying screams, a boulder smashed against the wall close to where we were standing. It crashed and splintered, jolting me out of my trance. In control of myself again, instinctively I dropped one of the flashlights to the ground, knowing it would continue to attract the attention of whatever was there, and leaped to the left. Oswaldo, still stunned, did not follow me. I shouted to him, but he didn't move. Then he suddenly reacted; he pulled his gun out of his belt and with a yell started shooting into the darkness. The sounds of screaming and gunshots echoed through the cave as more boulders spun through the air and smashed against the wall. All the while Oswaldo kept shouting and stopped firing only when his gun was empty.
>
> The howls continued as we fled, retracing our steps to the sharp bend in the passageway. But before we rounded it, I looked back toward the light left behind on the ground while trying to control my breathing and my heart's furious pounding. A tall, lumbering silhouette flashed across the shaft of light.

> The howls now became more intense, reverberating through the cave, and Oswaldo and I began to run toward the faint light of the entrance. Splashing through the pools of water, losing our footing, stumbling over the rocks and rubble strewn in our path. At last—Oswaldo first, with me close behind—we reached the opening, grabbed our packs, and scrambled over the debris to the outside.
>
> There I stopped and looked back. The roars and ear-splitting howls continued from within, then suddenly, silence.[2]

This chilling account may or may not have been a South American Sasquatch-type. Based on my personal experience with a plethora of encounters over the years, I have found the hairy folks to be friendly—even helpful at times, never aggressive. Because they can communicate, at times, with sophistication, I cannot help categorizing Turolla's creatures as animal-primates and not human-primates as the Sasquatch have proved to be. Of course, anyone might become aggressive, even violent, if someone simply walked into the privacy of his or her home, uninvited, carrying a gun! They had walked into the creature's lair and were driven out as one might expect. There could have been a mother bear or mother jaguar with cubs in the cave, so it could have been a lot worse.

Like the Peter Miele and Pino Turolla stories, more and more people are encountering global anthropoids as modern man continues to invade isolated regions of our planet. Not only are indigenous peoples losing access to their formal territories, but also wildlife—including Sasquatch-types everywhere—are presently being challenged by those from the outer world. The fact is that the creatures on this planet are running out of the living space needed to adequately survive. Additionally, toxic pollution is devastating every part of the food chain in previously pristine areas as well. It's a global natural disaster.

World traveler and biologist Steve Backshall says,

> Over the course of prehistory, there have been at least five great mass extinctions of wildlife on Earth. Each was due to a cataclysmic event, such as a meteor strike, volcanic eruption or dramatic climate change, wiping out anything up to 96% of the species then extant on the planet. Anyone who spends any time working with wild animals can be in no doubt that we are

now on the verge of a sixth mass extinction, and that this time it is we who are to blame.[3]

To add to this ecological chaos, most of the populace of third world countries is severely impoverished. In Asia, Africa, and Amazonia many people survive by poaching rare and exotic animals for body parts and for pets or private zoological gardens. This is a serious problem that is literally forcing some species to the edge of extinction. In the world's oceans alone, 75-80% of the fish populations have disappeared. The extraction of oil, water, and minerals has destabilized the Earth's geophysical structure as well. In humans, an increase in cancer and other diseases is a clear barometer of just how toxic our biosphere has become. If the future of our planet is at stake with all the living things on it, then where is the future of Sasquatchery? What will happen to the Sasquatch, Yeti, Yowie, Orang-pendek, Aluxes, Mapinguary, Mono Grande, and other hairy folks around the globe as environmental carnage continues to impact the food chain on every level? We are not meant to save the Sasquatch by stealing away the wilderness where they live and placing them on a "protected" reservation of land. The giants need to be where they are free-roaming and unmolested.

The future looks equally grim for the many remaining tribal peoples of the world—from the Amazon Basin, Africa's rainforests, Melanesia, and Asia. It is an anthropological fact that whenever a more dominant, technologically advanced society encounters a less technological people—the less technological suffer indignities—even genocide. The Peruvian government estimates they have 15 uncontacted tribes that are slowly being invaded by oil companies and illegal logging, according to Survival International. So this should give the reader an idea of the remoteness in Amazonia's 2,230,000 square miles of tropical rainforest that includes parts of Bolivia, Venezuela, Guyana, Colombia, Brazil, and Ecuador. It is estimated that in Brazil's interior there are up to 60 uncontacted tribes. The previously unknown Metykire tribe of 87 men, women, and children in western Brazil was attacked by White men with guns who killed 15 members. The Indians fled their village with the murderers in pursuit. They are now refugees, being exposed to contagious diseases. As an isolated people living in a jungle wilderness, where is their future—where are their lives headed? One might have hoped such invasive actions ended at Wounded Knee!

The Sasquatch and Ancient Ones want us to stop our moronic behavior and grow emotionally and spiritually mature, so genuine contact can be made. Telepathic communication is the key! They have told me and others that our societal values and actions are like those of children, and they feel like parents, watching the carnage and disrespect toward each other and the environment. For them, this is greatly disconcerting. The beings say we have a lot of intelligence and technology, but little wisdom in our society to apply it in a healthy way. Let's hope we are wise enough to avoid a possible all-out World War III!

The Sasquatch people say that since 2000 they are coming out more and allowing themselves to be seen worldwide. It's no coincidence, because these beings have a psychic network that can communicate to other races of their people anywhere on the globe. They want to sensitize the public and validate that they do indeed exist and are a part of our physical world. They have also repeatedly said that if man continues at the rate of destruction of the biosphere, with our dysfunctional socio-political behavior and blasé attitude toward ecological issues, then all races—including humanity—are *doomed.* To save the planet is to save ourselves as well as our Sasquatch neighbors. Tibet's Fourteenth Dalai Lama says about our living planet and its occupants, "If you see yourself in others, then who can you harm?"

The Sasquatch will help show us the way, if we would only seek out their wisdom from our hearts, not with a gun or just the intellect. This whole book is born of the Sasquatch's intention to help us. They are the ultimate environmentalists. These giants *are* the real wisdom-keepers of Mother Earth. They seem to have the ability to "read" or "tap into" the living Earth and its flora and fauna. They can process and synthesize the information on a sophisticated level of intellect way beyond the limits of modern science. The Sasquatch possess great wisdom that they are willing to share in order to avoid a tragic human-precipitated cataclysm. But are we ready?

Pushoma, the chief of the Ancient Ones in Oklahoma, indicates in this note that he asked Haloti to write that humanity has been at this juncture before:

Dream Man

This what Pushoma tell to write

Our people come here before time of Thropus

God send us here on ball of light

We build great city of stone on these mount

We are here to watch over what God creat

Nothing is new in your people creat

Many time Humans make machines and city

Them make war and destroy then start over

The future of cryptoanthropology and cryptozoology, or any other natural science is directly related to how well we can steward our living planet. If we are to succeed, indeed survive, in the 21st century, we need to quickly change our political protocol when attempting to meet new races of people. George Santayana wrote, "Those who cannot remember the past are condemned to repeat it."

Most ETs seem to concur with Pushoma. They too stand ready to help. However, we need to listen. They are our cosmic elder brothers. Their values and concerns are very different from those currently being expressed on Earth. Some of the books I have read about meeting a cosmic culture refer to artifacts, technology, trade, and exploration. We must put aside our fears, biases, and judgments; then we will have a far better chance to meet "star cultures" from other planets and those inhabiting other dimensions. This, of course, includes all the humanoid Sasquatch-types around the globe.

This is not something that can be left to our governments to implement. Perhaps I am being a pessimist, but I really don't think our world governments have the spiritual wisdom to make such a profound change. The evolved ETs on Planet Earth, who live in underground and underwater bases, listen in on secret government business, so they know our agenda! They can read our thoughts. For survival of the biosphere—including humanity—*healing thoughts must begin to come from our hearts.*

It is important for us to know that we are not lords of this Earth; we are its children and servants, who should act as planetary care-givers—not care-takers—if we are to keep a symbiotic relationship in balance. From our intense industrialization and exploitations, all the money and power we get is not worth the price we will ultimately pay. In Chief Seattle's 1854 presentation, he solemnly said,

> Teach your children what we have taught our children, that the Earth is our mother. Whatever befalls the Earth befalls the sons of the Earth. If men spit upon the ground, they spit upon themselves.

There is still a lot to explore on Planet Earth, if we can stay alive long enough to discover it. The Japanese have their unknown primate called the "Hibagon." More people are becoming aware of Sasquatch-types on several islands throughout the Pacific Ocean. Some may be non-human primates who evolved from a Darwinian perspective, and some may be hairy humanoids who were "seeded" on this planet, or migrated from a similar interdimensional world. We need them as allies to help us "run" the planet, to keep it healthy and in balance, since they live close to nature. I believe that our future is to learn to love and respect ourselves, whereupon we will have love and compassion to draw from to give to others whether they be friends, plants, animals, or exotic humanoid species. At present the future of Sasquatchery is tied to our own. Will we learn in time to create a healthy, peaceful world in which to dwell? When this is achieved, we won't need to look for Sasquatch for they will have already found us!

CHAPTER 10

THE SASQUATCH MESSAGE TO HUMANITY

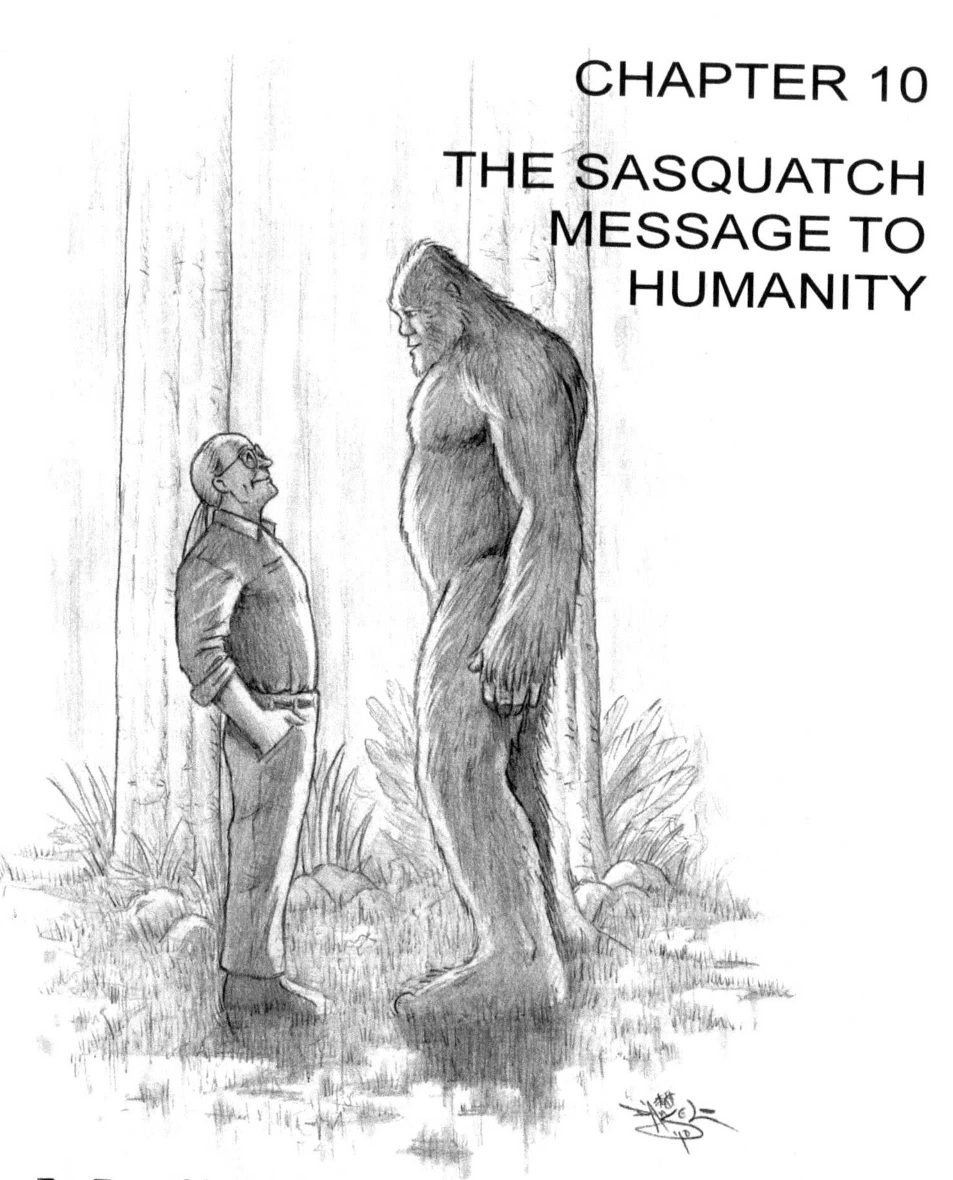

Many of the forest giants are willing to help people who are working for a better, healthier, more peaceful existence on Mother Earth. They emphasize that everyone must work together toward a common goal of sustainably sharing the air, water, soil, and food on our planet. However, some Sasquatch tribes still refuse to interact with us, because of our disrespect towards them and other inhabitants on the planet.

Modern people should also be working toward world peace and not settling our differences with war. The Sasquatch people say that World

War III has already begun, but has not yet culminated, so the populace has not fully experienced it yet.

Contactee Arla Williams of Oklahoma said that the Sasquatch council of elders has noted the following traits in human society:

A Sasquatch that was accidentally photographed in 1986 at the 55-acre Butchart Gardens near Victoria, Vancouver Island, British Columbia, Canada

1) we have created an illusion that blocks our awareness, so we don't see what the real focus in life is;
2) our control of others and nature while dominating living things around us is highly criticized by the giants;
3) this leads to a distrust of humans;
4) the nature people avoid us because of our "aberrant," destructive behavior;
5) we are too judgmental and bigoted, they emphasize;
6) we lack integrity.

This is why the Sasquatch choose only certain people to contact, whom they feel they can trust and who are devoted to serving the planet.

Close-up from previous page, of a Sasquatch on a walkway at Butchart Gardens.

Both ETs and Sasquatch view 2012 as the most important cosmic moment in human history, when our sun will be in perfect alignment with numerous other suns (stars) all the way to the very center of the Milky Way Galaxy, roughly 28,000 light years away. They said that the years 2007–2016 are the window during which these Earth changes are supposed to build up and occur—not exactly on 2012. Interestingly, these statements match up with Maya, Hopi, Nostradamus', and Bible prophecies. This event happens in our solar system and Milky Way Galaxy every 26,000 years. So present-day Earthlings really don't know what to expect during these times.

The Sasquatch people believe and are telling contactees that there will be an increase in solar flares, super-quakes, tsunamis, and possibly a pole flip, which has happened before. Super-storms, such as tornadoes and hurricanes, will begin to occur in greater frequency, and science has already confirmed this. The Sasquatch say the Earth's axis will move. Just how and to what degree is uncertain.

The giants said that all the recent earthquakes are loosening up the tectonic plates around the "ring of fire," which is a warning that major earthquakes will increase in number. When there is a shift in land mass in coastal regions, enormous tsunamis will follow, they have said. It is like a dog violently shaking off its fleas, and the Sasquatch insist that Mother Earth is truly a living organism with cosmic consciousness.

The giants say a series of global events will create a catastrophe that will destroy infrastructure of many countries, leading to war, chaos, famine, and infectious diseases. The Ancient Ones told me that one billion to a billion and a half people will be left. They told another contactee that one-quarter of humanity or less will survive. So there are many events that even they are uncertain of. Food, water, and medicine should be stored away, Haloti advised. Living in the country is safer than any city or large town. They are preparing for Earth changes and will take in some Earth humans when the shift reaches a danger point, they said. Staying close to God and selflessly helping others are important factors that will aid in survival, they also insist. Our preoccupation with materialism and fast-food dining has caused us to become out of touch with family and community. We must honor ourselves, they say, and learn self-respect. The meteorological calamity will force people back to basics, using ancient wisdom and more local agriculture.

A photograph of an Ancient One taken by a hiker in 1992 in the Sangre de Cristo Mountains, Colorado

They told one woman contactee that, during the time leading up to 2012, our DNA is slowly being restructured, If a person's spiritual psyche is being used for the highest good for all concerned, then that individual will be synchronized with the new DNA arrangement and the transition will be easier. The materialists will have a far more painful time of it, they said. The remnant or "junk" DNA medical science speaks of is supposed to be activated and rearranged. Such results will create more spiritual/psychic people more closely aligned to the Starpeople, Sasquatch and Ancient Ones, they said.

Only time will tell if the giants' message to humanity is correct. Some will prepare for it; most will not! The beings say that *love* is our greatest virtue at this time, and *fear*, its opposite, is our worst enemy. This is the message behind the Sasquatch phenomenon.

The First Nation people say that the Bigfoot are being seen with greater frequency, and that this is a sign that something important is about to happen; it's a warning. This shift is helping us to move away from our addiction to materialism, so we can be in better touch with

Track of an Ancient One in northeast Texas 2004

our soul-selves—our God-self—and start acting like a modern society. The forest beings want us to go beyond the limited material world of empiricism into a fully comprehensive understanding of Bigfoot, ETs, interdimensionalism, quantum mechanics, and psi. This will holistically transform our awareness into a type of "spiritual experiential model" encompassing a broader cosmographic spectrum than ever before. Then, the scientific model will be complete, representing the true, unlimited, more objective Universal Reality, which most have never seen before.

The Sasquatch tribes emphasize the necessity of this vital change in our mind-set if we are to survive the coming Earth changes. There are so many other life forms—Sasquatch, ETs, and other dimension-hoppers—all around us in the quantum soup of God's universe, trying to show us the way. The Bigfoot are spiritual keepers of Mother Earth and they will do whatever is needed to maintain her integrity, whether one believes in them or not. So the message from them is for us to drastically change the way we think about ourselves and the world around us, in order to produce a quantum shift in the global brain. It would be the *ultimate* achievement for humanity's survival and growth—recreating a "mature" civilization with more compassion, peace, spiritual awareness, and intellectual understanding. That is what the Sasquatch, et al., want for us so that we will raise ourselves closer to their level of consciousness. Ervin Laszlo says,

> When people evolve transpersonal consciousness they become aware of their deep ties to each other, to the biosphere, and to the cosmos. They develop greater empathy with people and cultures near and far and greater sensitivity to animals, plants and the entire biosphere. As a result a new civilization can see the light of day.[1]

This is precisely what the community of interdimensional beings is trying to convey to the human populace!

Laszlo continues,

> Achieving transpersonal consciousness is likely to further progress toward a civilization based on empathy, trust, and solidarity, a Holos-civilization. But will such a civilization come about in time? This we do not know yet. We do know that more people will achieve transpersonal consciousness in the coming years, and if we do not destroy our life-supporting environment and decimate our numbers, a critical mass may do so.[2]

We live in a fragile world where any number of negative acts or decisions could start a domino effect of destruction spiraling into extinction. This can be avoided if people rearrange their priorities, then become more *proactive*. Optimally, the archaic idea of monster hunting will end—replaced by meaningful dialogue with these sagacious people.

While on an expedition to Oklahoma, I asked Pushoma, the chief of the tribe of the Ancient Ones in that region, if he would "write" for people in the outer world the most meaningful message we need to hear. He agreed. This is what he wrote:

To Dream Man
For Book

No man is greater
or less than any other

The is but one God
the Father
the Creator

Do for others as you
wish to do for you

There in many spaces
does God dwell

This is but one reality
of untold many others

God Bless all
We are all part of the
whole

This is how Pushoma and his people think, and with his words, it appears that he is attempting to gently dissolve our anthropocentrism, because these basic statements are really a spiritual foundation for a more copacetic life. This should be our preoccupation as we approach 2012. Laszlo concludes:

> This quantum shift in the global brain is humanity's best chance. Margaret Mead said, "Never doubt the power of a small group of people to change the world. Nothing else ever has." Small groups of people with an evolved consciousness will change the world—if they grow into a critical mass in time. There could not be a nobler or more important task in our day than to empower this evolution.[3]

Author while on an expedition in the Selma-Bitterroot Wilderness, Idaho

AFTERWORD I

BY JANN WEISS
INTERSPECIES COMMUNICATOR

I first heard about Kewaunee through a friend who heard him speak at a workshop in Hawaii. My ears perked up and I found myself saying, "I need to know more about this guy." When I saw the title of his first book, *The Psychic Sasquatch*, my response was, "Oh, my God, someone else knows about this, *and* they've written a book." What impressed me the most about the book was the "matter of fact" approach to the subject. There was no question about whether or not the Big Hairy Folk existed. There was a lot of confirmation of information that I had "received" and explanations for the things that I didn't yet understand. It was one of those pivotal books that somehow made it okay to believe and trust my own experiences.

I met Kewaunee a couple of years later in a quaint little restaurant in a small town outside of Seattle. He came across as a no-nonsense, down-to-earth kind of guy. Not what one would expect of someone who spends a rather large amount of time hanging out in wooded areas communicating with a life form that is completely outside the reality of most humans. It didn't take very long at all, however, to see that he was for real. He wasn't trying to sell me on anything. He was simply willing to share with someone who would take him seriously; which I did, because I could "see" what he was talking about. It was as though I could tap into the minds of some of the beings that he'd been communicating with. Yes, I was definitely going to have to know more about this guy and his big hairy friends.

I first became aware of the existence of The Big Hairy Folk sometime in the '70s; I heard stories, saw pictures of plaster-of-paris casts of large feet and the occasional picture of a creature with long swinging arms. No one seemed to know what they were or if they were real—and if they were, what we should do about them. I was fascinated. Somehow, I knew that at least some of the stories were true, but I didn't know why I knew that.

During the '70s and '80s my own psychic ability was growing and expanding to include hands-on healing, channeling, and working with doctors as (what they now call) a medical intuitive. Information for or about other people just "showed up in my head" and when I shared that information, it seemed to make a difference in people's lives. Although I understood that every living human was intuitive to some degree, for me, it was as much a part of my life as my hearing or my eyesight. In 1987 it expanded again to include communication with another life form that shares our planet: dolphins. They were talking to me mind-to-mind, and they responded to my communications as well. And that was the start of my love affair with interspecies communication.

Over the years, I've been fortunate to have had the opportunity to share with and learn from a number of species. Whales have taught me that it's possible to be incredibly cosmic and extremely grounded in a physical body all at the same time. Elephants, who are very much like us in their perceptions and reactions to life's experiences, have shown great patience, trust, and a profound sorrow in knowing that our species is eliminating theirs from this planet. Chacma Baboons have abundantly demonstrated that humans are not the only primate evolving on Earth. And of course, the dolphins have shared their wisdom, their amazing sense of humor, and their ability to play while in physical form. I especially get a kick out of their ability to travel in and out of alternate realities for exploration and fun.

And then, there are the Big Hairy Folk. They are another species that shares our planet. They share many of our innate primate responses and, at the same time, are nothing like us. Like dolphins, they are capable of experiencing alternate realities. In fact, for most of them, *we* are in the alternate reality. I cannot explain how I know this to be true except to say that I have experienced it.

My own experiences with this remarkable species have included visual sightings, non-visual interaction (i.e., bumps and nudges), audible sounds after specifically asking for confirmation of their presence and telepathic communication. I've been fortunate enough to be the recipient of their healing abilities as well.

I have learned a lot from them, but am always a bit surprised when they ask to learn something from me. I remember sitting on a log watching friends "harvest" crystals in Arkansas one hot afternoon. A very large physical presence joined me. (Fortunately, I already knew he was nearby.) As we sat together watching my friends, he "mind-asked," "why do your kind take the bits of Earth?" I got the sense that he and his people were aware of the number of humans who dig up crystals at that particular mine. I explained to him that those shiny bits of earth had an effect on our bodies. They calmed us and helped our bodies to become more open to our Higher Power's influence and wisdom. I explained that when we took those shiny bits of earth home with us, it helped to make us better humans.

After thinking about my answer for awhile, he responded that it was very good for my kind to find a way to become more peaceful and open to the wisdom that comes from "higher places." I got the distinct impression that he was actually very happy with the idea. Before he left, he also expressed his gratitude for having had the opportunity to communicate and interact with one of my kind in a way that was not stressful for him and his kind.

And this leads me to another subject, one that Kewaunee speaks about frequently. We often hear the question, "Well if they exist, why is it so difficult to find them when we go out looking for them?" There is a definite knack to getting the big hairy folk to interact with you. First, your intentions must be clear and non-aggressive. (How enthusiastically would *you* embrace the idea of conversing with someone who was looking for you with a gun in his or her hands?) Second, it helps to communicate your appreciation for the interaction. And third, it's important that you don't try to direct the interaction. I always start my communication with "we share the same Earth," which means "we are equal here." This is something the elephants taught me when I lived in South Africa. Keep in mind that this species, like every other intelligent species on this planet, has its own point of reference and way of doing things. Reality is quite different for them. Whether a researcher or curious human, it helps to approach the subject with an open heart, a willingness to be surprised, a connection with your own Higher Power (to reduce the body's fear), and a lack of needing them to do anything. They are very much like us, and nothing like us at all. They are not Gods. However, they do have a way of being that is worth knowing about. We share the same Earth. We can learn from each other.

In closing, I would like to say that I think the most important gift that Kewaunee gives us through his work and his willingness to share his experiences, is a kind of freedom to continue our own research/

adventure in the world of Sasquatch. And by freedom I mean; it's easier to "exist outside the box" when there are others out there, too. Especially when the "others" are educated, knowledgeable, down-to-earth people.

So I thank you, Kewaunee. Thank you for taking the risks and for speaking out in spite of the negative reactions that this work frequently creates. Thank you for sharing and for making it a little easier for us to have and to trust our own experiences.

Well met, Brother.

May we all, learn to open our hearts to that which is possible, to that which enlightens us, to that which expands us.

Jann Weiss
Interspecies Communicator
Dallas, Texas
April 2010

Jann Weiss, interspecies communicator

AFTERWORD II
BY KATHLEEN JONES
SPIRIT MEDICINE & INTERSPECIES COMMUNICATION

When Kewaunee asked me to write an afterword to his second book, I readily accepted his invitation because I believe that Kewaunee honors and loves the Sasquatch people for who they truly are. I know that he has literally devoted a lifetime to exploring and sharing his experiences, so that heart-full humans everywhere can expand their own knowledge and experience with these delightful and profound beings. I see him as a teacher to the many who know in their hearts the truth of the wise and ancient Sasquatch people.

I have shared my life with the Sasquatch people for as long as I can remember. My initial recollection of their living in our midst was as a young girl in the mid-1950s, spending summer vacations with my family at our family cabin not far from Mt. Lassen in the mountains east of Redding in Northern California. I recall these first experiences as eliciting at once childlike curiosity and at the same time a mature knowing about who these beings are, living gently in the forest and providing only very selective and fleeting glimpses of themselves. The instant familiarity and camaraderie I felt with them I now believe came from numerous past lifetimes of living either with the Sasquatch people or, in the least, nearby and in close contact with them. I have never felt fear of them; in fact, from the beginning my desire has been for closer, more frequent, more intimate contact with them. In those early days I recall their nightly howls, hoots, and banging about in the brush and through the forest near our cabin. I remember riding in the family station wagon down deserted logging roads with my father as the "tour guide," all the while feeling the presence of my forest friends nearby and around us. It may sound quirky, but I have always heard their howls and felt their presence *in my heart*. It was as if each howl was calling me home to a very familiar place. For me, there is always the feeling of no time, no space when in their presence. From my earliest experience, I have felt loving towards and protective of the beautiful Sasquatch beings (though I will not be so presumptuous to believe for one moment that they need my protection!)

From those summer evenings and pre-dawn mornings in the mountains of over five decades ago there have been many events all weaving together in the tapestry of my life. The Sasquatch family may not have been so nearly close later, as I completed my studies at the University of California at Berkeley and moved on into the business world and raising a family of my own. Still, there were summers at the family cabin, and the Sasquatch were always there to welcome me back into our continuing shared and magical experiences. After nearly three decades of working and living in rural Mendocino County, California, and then on the island of Oahu in Hawaii, I continued to have a firm *knowing* that the Sasquatch were never far removed from my daily life experiences. That *knowing* has always been a source of comfort to me, no matter where I have found myself. For years I was occupied more than full time with my career and raising a family, and so my experience with the Sasquatch people remained mostly static, but certainly not far from instant recall—putting a smile in my heart whenever they entered my consciousness. Though focused on a mainstream business career for two decades, I had as early as the 1970s begun to awaken to the intuitive telepathic side of myself. This awakening called me to remember that I AM myself spirit, connected on a spirit level to all beings no matter their physical manifestations. In 1996 I retired from my business career to devote full time to expanding my spirit path. I began to create a professional spirit medicine practice of intuitive counseling and interspecies telepathic communication. The ability to telepathically share communication with other species/beings, both seen and unseen, has taken a big step of expansion and deepening in the past dozen or so years for me. Specifically, my connection to the Sasquatch people became a primary focus of my spirit work when I moved from Oahu to a little valley in the northernmost reaches of the Siskiyou Mountains in southwestern Oregon in 1999. My telepathic connection to them deepened as I began to experience them daily, and not infrequently with physical form in this dimension. I have been honored and blessed to reside in their home here in the woods for the past eleven years.

For those of you who want or need physical "proof" of the existence of Sasquatch—in other words, to all of you "researchers" out there—don't bother to read further. I have never aspired, or studied, to be a scientist. And my words here are not intended to meet even the most minimal scientific requirements for "proof." The Sasquatch people are a people *of the heart*, and so am I. That is what I am writing, both from my heart and channeled from theirs. I have been blessed with the expanding ability to connect with the heart and spirit of most all species. The information I receive has been confirmed over and over, both in the hearts and minds—yes, minds—of many. That said, this writing is not about my qualifications or skills as an interspecies communicator; it's just who I am. I don't join Sasquatch research/study groups or forums—just to hear these beautiful people referred to as "Bigfoot" grieves my heart and soul. You doubters out there, who choose to continue dancing in your *heads*, may want to refer to this brief writing as "Sasquatch According to Kathleen." Those of you who have peeled away the layers of doubt in your heads of the truth of the Sasquatch people, and choose now to follow the beckoning of your hearts, will hopefully recognize the truth being shared in both Kewaunee's writing and in mine. The time for arguing and debate is now over; it is now time for *listening*.

So, I will tell you, if you don't already know, that the Sasquatch family *is* humanoid, just like us. They are kind, gentle, intelligent, playful, and oh so very wise. They are people! They are not an *animal* curiosity to be exploited and hunted for the sake of glorifying the human ego/mind. They are our brothers and sisters! I choose to see them as our older, more ancient, brothers and sisters. Yes, we can mate with them. Unfortunately, human consciousness has evolved (perhaps de-volved?) to the point where it is difficult for most to recognize and accept that there may be a humanoid species on this planet that is older and wiser than we are. If contemporary (wo)men cannot analyze it with the mind, then their minds have convinced them it either isn't real or it certainly is of less intelligence and less value to the whole of life than we humans are!

This brings us precisely to the dilemma of why the Sasquatch appear to be so illusive to their younger, more impetuous human brothers and sisters. The Sasquatch do not intentionally show themselves to any human who has not at least begun to open his or her heart to relating to the world from a heart-full place. *The Sasquatch literally cannot hear you if you are not intending/speaking from your heart!* It really is that simple. *You cannot fool them.* That is why they know when you have a gun or camera, even if you think it is hidden from them. That is why they don't become captive in the multitudinous complex traps that continue to be set for them by the self-proclaimed brilliant researchers, scientists, and hunters. The Sasquatch get it, way more than even the most intelligent human researcher/scientist can imagine. So, as long as humans continue to focus on such childish, selfish foolishness, the Sasquatch will continue to elude them. *It is wholly an affair of the heart.* The Sasquatch are very patient and long-suffering with our tribe of humans. Their appearances and sharing of themselves are very intentional. There are no accidental *sightings*. On the other hand, the average Sasquatch researcher is all the time thinking (therefore intending), "How can my sighting benefit me?" rather than, "We're all in this together. How can my shared experience with Sasquatch benefit *all* species as we share the Earth together?"

The primary foundational focus of my work as an interspecies communicator and teacher is assisting those humans with a heart-full desire, in learning to focus and communicate with the heart rather than the ego/mind. There's a whole new world out "there" and in "here" for those humans choosing to walk and live in the new paradigm of *heart*! Until we recognize and *live* the fact that we're all made out of the same stuff, that we—all species—are divine Spirit (each with unique physical bodies), created to live together in peace on this beautiful planet Earth, we will founder in sickness, war, confusion, and all that befalls most life on the planet at this time. It doesn't matter how many books we have read, what degrees we hold, or the extent of our material accumulation. It is about our consciousness of life, how we live, how we honor ourselves and others (*all* species, *all* life). This perspective could easily be classified as New Age "woo-woo" by many choosing to channel their lives through their minds rather than their hearts. However, there is nothing "new" about the human recognition of the importance of honoring, rather than exploiting, *life* in all forms. In the times we are facing now, good intelligence and good learning will be absolutely essential *only when* they are channeled through our hearts; alone, minus the heart channel, these two qualities/achievements are quickly becoming obsolete and useless.

We have much to learn about this alternative perspective from the indigenous peoples of the planet. Specifically but not solely, most Native American tribes share similar lore and experiences of the human connection as the Sasquatch people—the importance of honoring one another and all of life. For many tribal peoples, the connection with Sasquatch was a given in daily life—naturally and normally. For many, the Sasquatch have been honored as especially wise teachers and healers.

In my own contemporary experience, this is how I have found them to be. It is from this perspective that I grow and cherish my connection with the Sasquatch living here with me and with those living in more distant parts of the planet, with whom I have been blessed to connect telepathically. There is nothing to fear about the Sasquatch people. They are loving, kind, and very family-focused. They easily illustrate the familiar motto, "Live and let live." But there is more to their purpose in being here with us than simply coexisting peacefully. We humans have brought ourselves, and other species to some extremely challenging and confusing times on the planet. It is no accident that the Sasquatch people are showing themselves and connecting telepathically with many, many more humans. It has nothing to do with the numerous and frequently appearing new books on "Bigfoot research." It has everything to do with the wisdom possessed by the Sasquatch and desperately needed by contemporary human civilization. The Sasquatch people have a significant role in shepherding the immense waves of light now coming forth across the Earth to raise the vibratory rate of human consciousness. It is time for us to recognize them, to honor them, to listen to them, to heed their wise instruction! If and when we inquire with our hearts, they have infinite wisdom to share with us younger brothers and sisters about reversing the destruction of the environment, the evolution of spiritual consciousness, how to live in peace and harmony. And as we grow in integrity of trust, honor, and heart-fullness, they will even teach us how to physically materialize and de-materialize at will as we walk through these rapidly changing times. And, yes, they do possess quite an enchanting sense of humor, so plan on some smiles and laughter as you include them in your journey. Those who have experienced physical connection with them will easily confirm the almost trickster nature they frequently express around us humans. They are similar to an older brother/sister who likes to tease a younger sibling.

In my spirit medicine practice that is focused around the Sasquatch people, I concentrate on 1) sharing techniques and assisting people to expand their heart-fullness, so that they can more directly experience the rich wisdom and joy of the Sasquatch, and 2) assisting the Sasquatch people in communicating and sharing this wisdom and experience with those humans who are just beginning to awaken to the new paradigm of peace, joy, wisdom, and enlightenment dawning on the planet. I do this by connecting my heart in sacred space to the Spirit of Sasquatch and then channeling whatever information/guidance/wisdom it is that they wish to share at the time. A telepathic communication session with them goes something like this (*for the sake of brevity in this writing, these sessions are quite abridged):*

Where are you as a people originally from?

Sasquatch: We are happy you are writing this. Our origins are long ago. We now consider Earth to be our home, though we have been known to re-visit our original homes in the stars. Some of us came from the Pleiades and some of us came from Arcturus. We came to Earth long before you humans appeared. We assisted in your birthing process. It was an intentional process and you are just now coming into who you were intended to be. It is almost like you humans are finally coming home. You will not look back to the eons of almost unconscious existence. You will soon see that we are of the same spirit. It is all the same. We are here to make the connection for you. It is happening now so quickly that it is almost electric, as you would say. We will put the plug in the socket for you. But many must recognize the plug, the socket before that can occur. The old Ones and some of the new Ones already know. You must open your hearts to them, and to us. [I got *old* Ones as a referral to the native indigenous people of the world and *new* Ones as a referral to the new group of souls arriving, also called *Indigo children* or *Rainbow children.*]

What do you see as we approach 2012?

Sasquatch: There will be a bright light. The light will be both inside and outside. There will be a burning off of the dark. It is time for the darkness to leave. The preparation is happening now and it is happening quickly. Those clinging to the darkness must let go. If they do not let go, they will be gone in the flash of light. There is a new way and we will all be there. You have been told by many about the promise. It will be easy for those in the light, impossible for those in the dark. There will be great change and it will happen quickly. The change is good. It has been told for many years. The change has already begun. We are connectors for you humans still walking step by step on the Earth. We will show you how to be faster than you could ever imagine. It is all here now, but most cannot see it and those who do only see glimpses. The light is too bright to see fully at this time. [They will be teaching us about what we humans call astral traveling, materialization/dematerialization, telepathic communication, etc.]

What do you have to say about the environmental changes occurring on the planet?

Sasquatch: It has been difficult for all of us. No species has been spared. The greed and fear must transmute or there will be complete destruction. This is a pivotal time. When the human consciousness enlightens, then the reversal of destruction will begin. We have specific important ideas that you humans need to learn and live. Just ask us and we will help you in ways you could have never imagined. It is critical to submit your intelligence to your Spirit.

What is your connection to the ETs?

Sasquatch: There are many, as you call them, "ETs." Some live in the dark and some live in the light. We came from the light. The "ETs" are more ancient of our brothers/sisters than you are. They are also your brothers/sisters and we are working together for the re-creation of the original intention. We have always assisted the "ETs" in their work here. They have a great focus on protecting you from yourselves. You might say that we are the guardians of the Earth Spirit and they are the guardians of the Sky Spirit. We must all work together for the light to be full. You humans are joining us.

Do you have any comments about this, Kewaunee's second book?

Sasquatch: He has always been with us. He knows us well. He is learning of the heart also. This new writing is a connective step for us. It is necessary in the process of the humans "seeing" who we are. He has always honored us because he has lived with us and we touched his heart many years ago. He has not forgotten. It has been a lonely journey for him. We know it has been difficult at times. We always care for him and protect him from those who do not understand. He helps others to understand. Sometimes he trips over his own rocks, but he knows we will always help him to walk forward once again. He is a most special brother to us.

My heart always feels so expanded and light (with love and joy) both during and after a telepathic session with the beautiful Sasquatch people. I am honored and delighted to facilitate telepathic sessions for clients and any who have chosen to vibrate from their heart perspective. Though sacred space shared with the Sasquatch often brings smiles and laughter, I take very seriously my Spirit connection with them and consider myself to be especially blessed to share their wisdom and love with the world.

Kathleen Jones
Spirit Medicine and
Interspecies Communication
P. O. Box 1737
Jacksonville, Oregon 97530
March 2010

sasquatchspeaks@terragon.com
www.sasquatchspeaks.com

END NOTES

Chapter 1

1. Sanderson, Ivan T., *Abominable Snowmen: Legend Come to Life*. New York: Chilton Book Company, 1961, page 68.
2. Sleigh, Daphne, *The People of the Harrison*. Abbotsford, British Columbia: Abbotsford Printing, 1990, page 83.

Chapter 2

1. Steenburg, Thomas, *In Search of Giants*. Surrey, British Columbia: Hancock House Publishers, 2000, page 116.

Chapter 3

1. Powell, Thom, *The Locals: A Contemporary Investigation of the Bigfoot/Sasquatch Phenomenon*. Blaine, Washington: Hancock House Publishers, 2003, pages 203–204.
2. Quotes from a letter, 1969.
3. Gordon, Stan, *Silent Invasion: The Pennsylvania UFO-Bigfoot Casebook*. Stan Gordon Productions, 2010, pages 227–244.
4. Wolverton, Keith, *Mystery Stalks the Prairie*. T.H.A.R. Institute, 1976.
5. Howe, Linda Moulton, *Glimpses of Other Realities, Volume II*. Jamison, Pennsylvania: LMH Productions, 1998, pages 145–157.
6. Johnson, Paul G., and Joan Jeffers, *The Pennsylvania Bigfoot*. 1980.
7. Guttilla, Peter, *The Bigfoot Files*. Santa Barbara, California: Timeless Voyager Press, 2003, page 215.
8. Dongo, Tom, and Linda Bradshaw, *Merging Dimensions*. Sedona, Arizona: Hummingbird Publishing, 1995, page 97.
9. White Song Eagle, *Teluke: A Big Foot Account*. Bloomington, Indiana: Authorhouse, 2008.
10. Kelleher, Colin A., and George Knapp, *Hunt For the Skinwalker*. New York: Paraview, 2005, pages 148–158.

11. Macer-Story, Eugenia, "Coming Back into the Circle: An Interview with Australian Medicine Woman Lorraine Malfi Williams" *Magical Blend*, June 1988, page 48.
12. Boirayon, Marius, *Solomon Islands' Mysteries: Accounts of Giants and UFOs in the Solomon Islands. www.solomonislandsmysteries.com*, 2009.

Chapter 4

1. Beck, Fred, *I Fought the Apemen of Mt. St. Helens*. New Westminster, British Columbia: Pyramid Publications, 1967, page 7.
2. Ibid., page 7.
3. Ibid., page 7.
4. Ibid., page 8.
5. Ibid., pages 9 and 10.
6. Ibid., page 15.
7. *Mail Online from the United Kingdom: www.dailymail.co.uk/news/article-490669/Army-tests-James-Bond-style-tank-invisible.html*
8. *www.iop.org*
9. Radin, Dean, *Entangled Minds: Extrasensory Experiences in a Quantum Reality*. New York: Paraview, 2006, page 3.
10. Imbrogno, Philip, *Interdimensional Universe: The New Science of UFOs, Paranormal Phenomena and Other-dimensional Beings*. Woodbury, Minnesota: Llewellyn Publications, 2008, pages ix–x.

Chapter 5

1. Steiger, Brad, *The Awful Thing in the Attic*. Lakeview, Minnesota: Galde Press, Inc., 1995, page 67.
2. Godfrey, Linda S., *Hunting the American Werewolf.* Madison, Wisconsin: Trails Books, 2006, pages 167–178.

3. Dongo, Tom, and Linda Bradshaw, *Merging Dimensions: The Opening Portals of Sedona*. Sedona, Arizona: Hummingbird Publishing, 1995, Page 75.
4. Guttilla, Peter, *The Diane Vaughan Story*. *www.bigfootencounters.com*, May 17, 1989.
5. Crowe, Ray, *The Track Record, Summary #158*. February 2006, page 10.
6. *www.theregister.com/uk/2009/11/06*
7. *www.sciencedaily.com/releases/2009/08/090813083329.htm*

Chapter 6

1. Ingraham, E.S., *The Pacific Forest Reserve and Mt. Rainier... A Souvenir*. Seattle, 1895.
2. Gilroy, Rex, *Mysterious Australia*. Kempton, Illinois: Adventures Unlimited/Nexus Publishing, 1st edition, 1995, page 199.
3. Swanson, Claude, *The Synchronized Universe: New Science of the Paranormal*. Tucson, AZ: Poseidia Press, 2003.
4. Laszlo, Ervin, *Quantum Shift in the Global Brain: How the New Scientific Reality Can Change Us and Our World*. Rochester, Vermont: Inner Traditions, 2008, page 84.
5. Ibid., page 85.
6. Ibid., page 122.

Chapter 7

1. Watson, Lyle, *Supernature*. Garden City, New Jersey: Anchor Press, 1973, pages 3–4.
2. Suttles, Wayne, "On the Cultural Track of the Sasquatch," *The Scientist Looks At the Sasquatch*. Moscow, Idaho: University Press, 1979, page 50.
3. Dennet, Preston, *Supernatural California*. Atglen, Pennsylvania: Schiffen Publishing, 2006, page 60.

4. Strain, Kathy Moskowitz, *Giants, Cannibals, and Monsters: Bigfoot in Native Culture*. Blaine, Washington: Hancock House Publishers, 2008, page 202.
5. Denny, Chaska, *www.bigfootsightings.org*, June 21, 2007
6. Ibid., July 16, 2007.
7. Deloria, Jr., Vine, *Red Earth, White Lies*. Golden, Colorado: Fulcrum Publishing, 1997, page 7.
8. Ibid., Page 7–8.
9. Goldenberg, Linda, *Little People and a Lost World: An Anthropological Mystery*. Minneapolis, Minnesota: Twenty-First Century Books, 2007, page 14.
10. Boatman, John, "The Star People," *My Elders Taught Me*. Milwaukee, Wisconsin: University Press, 1991, pages 93–102.

Chapter 8

1. Cremo, Michael A., *Forbidden Archeology: The Hidden History of the Human Race.* San Diego: Govardham Publishing, 1993, page 150.
2. Kannenberg, Ida, and Lee Trippett, *My Brother is a Hairy Man*. London: Experiences Enterprises, 2009, page 196.
3. Powell, Thom, *The Locals: A Contemporary Investigation of the Bigfoot/Sasquatch Phenomenon*. Blaine, Washington: Hancock House Publications, 2003, page 7.
4. Coleman, Loren, *Men in Cryptozoology: Daniel Perez. www.cryptomundo.com/cryptozoo-news/perez-mic*, May 31, 2008.
5. Matthew, Rupert, *Sasquatch: True-Life Encounters With Legendary Ape-Man*. London: Arcturus Publishing Limited, 2008, page 113.
6. Ibid., page 114.
7. Scurlock-Durana, Suzanne, *Full Body Presence: Learning to Listen to Your Body's Wisdom*. Novato, California: New World Library, 2010.

Chapter 9

1. Turolla, Pino, *Beyond the Andes*. New York: Harper & Row, 1980, page 135.
2. Ibid, pages 286–288.
3. Backshall, Steve, *Venomous Animals of the World.* United Kingdom: New Holland Publishers, Ltd., 2007, page 84.

Chapter 10

1. Laszlo, Ervin, *Quantum Shift in the Global Brain: How the New Scientific Reality Can Change Us and Our World.* Rochester, Vermont: Inner Traditions, 2008, page 125.
2. Ibid., page 125.
3. Ibid., page 126.

BIBLIOGRAPHY

Alley, Robert J., *Raincoast Sasquatch.* Blaine, Washington: Hancock House Publishers, 2003.

Arment, Chad, *Boss Snakes: Stories and Sightings of Giant Snakes in North America.* Landisville, Pennsylvania: Coachwhip Publications, 2008.

Ash, David, and Peter Hewitt, *The Vortex: Key to Future Science.* Bath, United Kingdom: Gateway Books, 1990.

Backshall, Steve, *Venomous Animals of the World.* Baltimore: The Johns Hopkins University Press, 2007.

Bartholomew, Robert E., and Paul B. Bartholomew, *Bigfoot: Encounters in New York and New England.* Blaine, Washington: Hancock House Publishers, 2008.

Beck, Fred, *I Fought the Apemen of Mt. St. Helens*. New Westminster, British Columbia: Pyramid Publications, 1967.

Bille, Matthews A., *Rumors of Existence*. Blaine, Washington: Hancock House Publishers, 1995.

———, *Shadows of Existence, Discoveries and Speculations in Zoology*. Blaine, Washington: Hancock House Publishers, 2006.

Boirayon, Marious, *Solomon Islands Mysteries: Accounts of Giants and UFOs in the Solomon Islands, www.solomonislandmysteries.com*, 2009.

Boatman, John, "The Star People," *My Elders Taught Me*. Milwaukee, Wisconsin: University Press, 1991.

Boone, J. Allen, *Kinship With All Life.* San Francisco, California: Harper San Francisco, 1954.

Brunke, Dawn Baumann, *Animal Voices: Telepathic Communication in the Web of Life.* Rochester, Vermont: Bear and Company, 2002.

Coleman, Loren, *Mysterious America.* New York: Paraview, 2001.

Corwin, Jeff, *100 Heartbeats: The Race To Save Earth's Most Endangered Species*. New York: Rodale, Inc., 2009.

Cremo, Michael A., and Richard L. Thompson, *Forbidden Archeology: The Hidden History of the Human Race.* San Diego: Govardham Hill Publishing, 1993.

Davenport, Marc, *Visitors From Time: The Secret of the UFOs.* Tigard, Oregon: Wild Flower Press, 1992.

Deloria, Jr., Vine, *Red Earth, White Lies: Native Americans and the Myth of Scientific Fact.* Golden, Colorado: Fulcrum Publishing, 1997.

Dennet, Preston, *Supernatural California.* Atglen, Pennsylvania: Schiffen Publishing, 2006.

Dongo, Tom, and Linda Bradshaw, *Merging Dimensions: The Opening Portals of Sedona.* Sedona, Arizona: Hummingbird Publishing 1995.

Dyer, Wayne W., *The Power of Intention.* Carlsbad, California: Hay House, 2005.

Franzoni III, Henry J., *In the Spirit of Seatco: Sasquatch, Indians, Geography, and Science in the Nineteenth Century.* Deer Island, Oregon: Ste Ye Hak Publishing, 2008.

George, Chief Dan, *The Best of Chief Dan George*. Surrey, British Columbia: Hancock House Publishers, 2003.

Gibbon, William J., *Mokele-Mbembe*: *Mystery Beast of the Congo Basin.* Landisville, Pennsylvania: Coachwhip Publications, 2010.

Gilroy, Rex, *Mysterious Australia.* Mapleton, Queensland, Australia: NEXUS Publishing, 1995.

Gittelson, Bernard, *Intangible Evidence.* New York: Simon and Schuster, 1987.

Godfrey, Linda S., *Hunting the American Werewolf.* Madison, Wisconsin: Trails Books, 2006.

Goldenberg, Linda, *Little People and a Lost World: An Anthropological Mystery.* Minneapolis, Minnesota: Twenty-First Century Books, 2007.

Gordon, Stan, *Silent Invasion: The Pennsylvania UFO-Bigfoot Casebook.* Stan Gordon Productions, 2010.

Gurney, Carol, *The Language of Animals: 7 Steps to Communicating With Animals.* New York: Bantum-Dell Publishing, 2001.

Guttilla, Peter, *The Bigfoot Files.* Santa Barbara: Timeless Voyager Press, 2003.

Harris, Paola Leopizzi, *Exopolitics: How Does One Speak to a Ball of Light?* Bloomington, Indiana: Author House, 2007.

Hart, John, *Hiking the Bigfoot Country: Exploring the Wildlands of Northern California and Southern Oregon.* San Francisco: The Sierra Club, 1975.

Hart, Will, *The Genesis Race: Our Extraterrestrial DNA and the True Origin of the Species.* Rochester, Vermont: Bear and Company, 2003.

Healy, Tony, and Paul Cropper, *The Yowie: In Search of Australia's Bigfoot.* San Antonio, Texas: Anomalist Books, 2006.

Holiday, F.W., *Creatures From the Inner Sphere.* New York: Popular Library, 1973.

Howe, Linda Moulton, *Glimpses of Other Realities, Volume II: High Strangeness.* Cheyenne, Wyoming: Pioneer Printing, 1998.

———, Linda Moulton, *Glimpses of Other Realities, Volume II.* Jamison, Pennsylvania: LMH Productions, 1998.

Human Potential Foundation, *When Cosmic Cultures Meet.* Falls Church, Virginia, 1995.

Hurley, Matthew, *The Alien Chronicles: Compelling Evidence for UFOs and Extraterrestrial Encounter in Art and Texts Since Ancient Time.* Chester, United Kingdom: Quester Publications, 2003.

Imbrogno, Philip, *Interdimensional Universe: The New Science of UFOs, Paranormal Phenomena and Other Dimensional Beings.* Woodbury, Minnesota: Llewellyn Publications, 2008.

Ingraham, E.S., *The Pacific Forest Reserve and Mt. Rainier...A Souvenir.* Seattle, 1895.

Johnson, Paul G., and Joan Jeffers, *The Pennsylvania Bigfoot.* 1980.

Jones, Marie, *PSIence: How New Discoveries in Quantum Physics and New Science May Explain the Existence of Paranormal Phenomena.* Franklin Lakes, New Jersey: New Page Books, 2007.

Kaku, Michio, *Parallel Worlds: A Journey Through Creation, Higher Dimensions, and the Future of the Cosmos.* New York: Anchor Books, 2005.

———, *Physics of the Impossible: A Scientific Exploration into the World of Phasers, Force Fields, Teleportation, and Time Travel.* New York: Doubleday, 2008.

Kannenberg, Ida M., and Lee Trippett, *My Brother is a Hairy Man.* London: Experiencers Enterprises, 2009.

Keel, John A., *The Complete Guide to Mysterious Beings.* New York: Doubleday, 1970.

Kelleher, Colin A., *The Complete Guide to Mysterious Beings*. New York: Doubleday, 1970.

———, and George Knapp, *Hunt for the Skinwalker*. New York: Paraview, 2005.

Kelly, Penny, *Robes: A Book of Coming Changes.*

Laszlo, Ervin, *Quantum Shift in the Global Brain: How the New Scientific Reality Can Change Us and Our World.* Rochester, Vermont: Inner Traditions, 2008.

Long, William J., *How Animals Talk and Other Pleasant Studies of Birds and Beasts.* Rochester, Vermont: Bear and Company, 2005.

Lydecker, Beatrice, *What the Animals Tell Me*. San Francisco: Harper & Row Publishers, 1977.

MacGregor, Rob, and Bruce Gernon, *The Fog: A Never Before Published Theory of the Bermuda Triangle Phenomenon.* Woodbury, Minnesota: Llewellyn Publications, 2005.

Malkowski, Edward F., *Sons of God, Daughters of Men.* Champaign, Illinois: Bits of Sunshine Publishing Company, 2004.

Marvis, Jim, *PSI Spies: The True Story of America's Psychic Warfare Program.* Franklin Lakes, New Jersey: New Page Books, 2007.

Maruyama, Magoroh, and Arthur Harkins, *Culture Beyond and the Earth: The Role in Anthropology in Outer Space*. New York: Vantage Books, 1975.

Matthews, Rupert, *Sasquatch: True-Life Encounters With Legendary Ape-Men.* Secaucus, NJ: Chartwell Books, Inc, 2009.

McKnight, Rosalind A., *Cosmic Journeys: My Out-of-Body Explorations with Robert A. Monroe.* Charlottesville, Virgina: Hampton Roads Publishing Company, Inc., 1999.

Meier, "Billy" Eduard Albert, *Through Space and Time.* Tulsa, Oklahoma: Steelmark, LLC, 2006.

Meldrum, Jeff, *Sasquatch: Legend Meets Science.* New York: Tom Doherty Associates, LLC, 2006.

Messner, Reinhold, *My Quest for the Yeti.* New York: St. Martin's Griffin, 1998.

Montgomery, Sy, *Working with the Great Apes.* New York: Houghton Mifflin Company, 1991.

Mott, Wm. Michael, *Caverns, Cauldrons, and Concealed Creatures: A Study of Subterranean Mysteries in History, Folklore, and Myth.* Frankston, Texas: TGS-Hidden Mysteries, 2002.

Nunnelly, B.M., *Mysterious Kentucky.* Decatur, Illinois: Whitechapel Press, 2007.

O'Brien, Christopher, *The Mysterious Valley.* New York: St. Martin's Press, 1996.

Paulides, David, *The Hoopa Project: Bigfoot Encounters in California.* Blaine, Washington: Hancock House Publishers, 2008.

———, *Tribal Bigfoot.* Blaine, Washington: Hancock House Publishers, 2009.

Powell, Thom, *The Locals: A Contemporary Investigation of the Bigfoot/Sasquatch Phenomenon.* Blaine, Washington: Hancock House Publishers, 2003.

Quinn, Ron, *Little People.* Lakeville, Minnesota: Galde Press, Inc., 2006.

Radin, Dean, *Entangled Minds: Extrasensory Experiences in a Quantum Reality.* New York: Paraview, 2006.

Raynes, Brent, *Visitors From Hidden Realms.* Memphis, Tennessee: Eagle Wing Books, Inc., 2004.

Red Star, Nancy, *Legends of the Star Ancestors: Stories of Extraterrestrial Contact From Wisdom Keepers Around the World.* Rochester, Vermont: Bear and Company, 2002.

Riggs, Rob, *In the Big Thicket: On the Trail of the Wild Man.* NewYork: Paraview Press, 2001.

Sanderson, Ivan T., *Abominable Snowmen: Legend Come to Life*. New York: Chilton Book Company, 1961.

Satchidananda, Sri Swami, *The Yoga Sutras of Patanjali*. Integral Yoga Publication, 1978.

Schellhorn, G. Cope, *Extraterrestrials in Biblical Prophecy.* Madison, Wisconsin: Horus House Press, 1989.

Scurlock-Durana, Suzanne, *Full Body Presence: Learning to Listen to Your Body's Wisdom*. Novato, California: New World Library, 2010.

Sheppard-Wolford, Sali, *Valley of the Skookum.* Pine Woods Press, 2006.

Sitchin, Zecharia, *Divine Encounters.* New York: Avon Books, 1995.

———, *There Were Giants Upon the Earth: Gods, Demigods, and Human Ancestry: The Evidence of Alien DNA.* Rochester, Vermont: Bear and Company, 2010.

Sleigh, Daphne, *The People of the Harrison*. Abbotsford, British Columbia: Abbotsford Printing, 1990.

Steenburg, Thomas, *In Search of Giants: Bigfoot Sasquatch Encounter*. Surrey, British Columbia: Hancock House Publishers, 2000.

Steiger, Brad, *The Awful Thing in the Attic.* Lakeville, Minnesota: Galde Press, Inc., 1995.

Strain, Kathy Moskowitz, *Giants, Cannibals, and Monsters: Bigfoot in Native Culture.* Blaine, Washington: Hancock House Publishers, 2008.

Suttles, Wayne, "On the Cultural Track of the Sasquatch," *The Scientist Looks At the Sasquatch*. Moscow, Idaho: University Press, 1979.

Sweet, Leonore, *How to Photograph the Paranormal.* Charlottesville, Virginia: Hampton Roads Publishing Company, Inc., 2005.

Swerdlow, Stewart A., *Blue Blood, True Blood: Conflict and Creation.* Saint Joseph, Michigan: Expansions Publishing Company, Inc., 2002.

Thompson, Richard L., *Alien Identities.* Alachua, Florida: Govardhan Hill, Inc., 1993.

Turolla, Pino, *Beyond the Andes.* New York: Harper & Row, 1980.

Watson, Lyle, *Supernature*. Garden City, New Jersey: Anchor Press, 1973.

White Song Eagle, *Teluke: A Bigfoot Account.* Bloomington, Indiana: Authorhouse, 2008.

Wolverton, Keith, *Mystery Stalks the Prairie*. T.H.A.R. Institute, 1976.

INDEX

ABOUT THE AUTHOR

Kewaunee Lapseritis, AS, BA, MS, is a Holistic Health Consultant, Master Herbalist and Master Dowser with background in anthropology, psychology, conservation, and holistic health. His academic degrees are from: Greenfield Community College, Greenfield, Massachusetts; University of New Hampshire—Durham; North Adams State College, in North Adams, Massachusetts, and University of Wisconsin—Milwaukee. He also attended Williams College in Williamstown, Massachusetts. Kewaunee taught anthropology for one year at North Adams State College and later taught dowsing and psychic development at the Wisconsin Society for Psychic Reasearch—Milwaukee. As a world authority on the Bigfoot/Sasquatch phenomenon, he has meticulously researched the subject for the last 55 years. Mr. Lapseritis is also a social scientist and both a Sasquatch and ET "contactee."

As a world traveler, an amateur naturalist, and avid backpacker with wilderness skills, Mr. Lapseritis spent five years out of the country living in England, East Africa, and Japan, later immigrating to Australia, and has visited over 40 countries around the globe. He was in the Himalayas in 1968 investigating the Yeti, or "abominable snowman." In 1973, he conducted an ethnographic study in conjunction with the Colombian Institute of Anthropology in Bogota, living amongst the Tukuna Indians of Upper Amazonia.

In 1979, scientist Kewaunee Lapseritis was first telepathically contacted by a Sasquatch and an ET simultaneously, which was the shock of his life! To further complicate matters, the contact changed him and he developed psychic ability overnight, which triggered a spiritual transformation. At the time, he was assistant director of an urban Indian agency, and had been working as a hypnotherapist part-time, as well as lecturing at the Medical College of Wisconsin—a background that left him ill-prepared for such a happening.

Kewaunee has been a guest on over 350 radio and television talk shows. To date, he has been featured in 23 books and is often featured in the international press, including various newspapers. Some of the publications are: Australia's *People and Post* magazines, *OMNI* magazine, *Magical Blend, Cryptozoology* (journal), *Wildfire, Fate, Argosy, UFO* magazine, *Health Consciousness*, and many others. In 1991, he was on a panel of scientists on a two-hour Bigfoot documentary on national television. Plus, he has appeared on The Discovery Channel twice. He has lectured and presented papers throughout the United States 70 different times.

There are people all over the globe encountering Bigfoot and UFOs simultaneously while the beings share information about themselves, human history, and the future of our planet. If anyone wishes to share in confidence their telepathic conversation(s) and experiences with the Sasquatch people, please feel free to contact this author and he will answer all correspondence personally.

Please send all correspondence to:
Sasquatchpeople@hotmail.com
Or call 425-844-8409

www.Sasquatchpeople.com

What appears as a monster;
What is called a monster;
What is recognized as a monster;
Exists within a human being himself
And disappears with him.

Milarepa, a Sherpa poet
(AD 1040-1123)
Himalaya Mountains

ABOUT THE ILLUSTRATOR

Illustrator Jesse D'Angelo is a native of New York, and a top storyboard artist and designer in the film and television industry. At an early age he became interested in art and storytelling, and at 16 began his professional career. His feature film credits include *Sky Captain and the World of Tomorrow, Darkness Falls, Hellboy, Species, Air Force One,* and many more. His television credits include *Cold Case, Without A Trace, CSI: New York*, and commercials for Sony, Mattel, Toyota, and others. He has directed several short films, written many screenplays and award-winning short stories, and is the author of the Photoshop graphic novel *Flight From Our Town.*

Fascinated by all things mysterious and unexplained since early childhood, D'Angelo's specific interest in Sasquatch sparked in 2005 when his research and artwork on the subject began. While he has never seen a Sasquatch, he combines artistry with science and uses cutting edge technology to create images that are as accurate as possible. His work is widely recognized in the field of cryptozoology, and he hopes to help in some small way to spread the truth about our enigmatic friends and put aside the stereotypes.

http://www.oregonbigfoot.com/artists/jesse_dangelo.php
http://www.famousframes.com/website/portfolio.php?user_id=71&user_type=2#